HUNTINGTON BEACH LIFEGUARD
BUD/S CLASS 47
UDT 13
VIETNAM VETERAN
APOLLO 13 CAPSULE RECOVERY
USC SIGMA CHI
1978 MOLOKAI HO' 1ST PLACE
LOS CABALLEROS
COMBAT MURAL ARTIST
FOUNDER OF PATRIOT SEAL HERITAGE FOUNDATION

GOLDEN
RAY
BOOK PUBLISHING

THE PAINTED FROG

PETE "THE PIRATE" CAROLAN

Foundation: Patriot SEAL Heritage Foundation 501 (c)(3)

PATRIOT SEAL
HERITAGE
FOUNDATION

GOLDEN RAY
BOOK PUBLISHING

Patriot SEAL Heritage Foundation 501. (c)(3)
Published by Golden Ray Publishing, LLC
Printed in the United States of America

ISBN Hardcover: 979-8-994-50801-5
ISBN Paperback: 979-8-9973736-0-3

COVER AND INTERIOR DESIGN: LAURA DUFFY DESIGN

Dedicated to the enduring memory of

Pete "The Pirate" Carolan

founder Patriot SEAL Heritage Foundation 501 (c)(3).

May the pages within serve as a beacon of hope, healing, and remembrance for those who continue to fight the silent battles brought home from fighting our nations wars.

This work honors our heroes living with the lasting effects of service, including PTSD, TBI, and related conditions, and stands in support of the men of our Special Forces, Navy SEALs, Vietnam Veterans, and active-duty and retired United States veterans, and first responders.

Proceeds from this book help support the mission of the *Patriot SEAL Heritage Foundation,* advancing critical programs, therapeutic equine retreats for Navy SEALS, and K9 training initiatives dedicated to restoring strength, purpose, and brotherhood.

For More Information or to connect for Media Inquiries and Art Installations, please visit the foundation at:

www.patriotsealheritage.org

Peter J. Carolan

Pete "The Pirate" Carolan

August 6, 1947 – August 27, 2022

THE PSHF FOUNDATION

Peter J. Carolan founded the *Patriot SEAL Heritage Foundation*. He wanted to honor the legacy of military service and inspire continued purpose through leadership, mentorship, and service beyond the uniform as a super patriot.

Created for the unsung heroes—those who have carried the weight of service without recognition—the foundation exists to restore purpose, structure, and brotherhood in the lives of Special Operations warriors, veterans, and first responders. Through strategic partnerships with aligned nonprofit and for-profit organizations, the foundation supports transformative programs rooted in PTSD, TBI, and related conditions—designed to address the unseen war many of our veterans fight when the uniform comes off.

Central to its mission is the powerful bond between human and animal. Through working dog programs, participants rebuild trust, discipline, confidence, and connection. Through therapeutic equine retreats on Catalina Island and wrap around therapy programs throughout Southern California, they discover a path to healing that does not require reliving trauma. Horses teach what words often cannot—presence, emotional regulation, awareness, and the ability to lead with calm under pressure. This is more than recovery. It is the return to purpose. We stand together to honor the legacy of "Pete the Pirate" Carolan and his mission for our warriors.

THE PAINTED FROG

PETE "THE PIRATE" CAROLAN

PETER J. CAROLAN

UDT 13 | COMBAT MURALIST | NATIONAL UDT-SEAL MUSEUM

MEDIA INQUIRIES: MEDIA@PATRIOTSEALHERITAGE.COM

NAVY-SEAL MUSEUM

PETER J. CAROLAN
BUDS CLASS 47 | UDT 13 | APOLLO 13 CAPSULE RECOVERY TEAM
COMBAT MURALIST
MEDIA INQUIRIES: MEDIA@PATRIOTSEALHERITAGE.COM

USS Midway MUSEUM
AMERICA'S LIVING SYMBOL OF FREEDOM

JOIN US IN MAKING A DIFFERENCE
WWW.PATRIOTSEALHERITAGE.ORG

FOREWORD

(RET.) USN COMMANDER SKIP CRANE

SEAL TEAM 1

Chairman of the Board - PSHF

The Artist of the Teams

My name is (Ret.) USN Commander Skip Crane. Today, I serve as Chairman of the Board of *The Patriot SEAL Heritage Foundation.* That title suggests stewardship, governance, and legacy. But long before I ever sat at a table discussing preservation or history, I was shaped by cold Pacific water, relentless training, and the humid jungles of Southeast Asia. I was shaped by men who learned to endure discomfort, accept risk, and place the needs of the team above their own.

To understand Pete Carolan, the man we called "Pete the Pirate," you must first understand the crucible that forged our brotherhood. And to understand why his art matters, you must understand what it means to fight two wars: one overseas, and the other inside your own skull. Pete fought both. One war had a beginning, an enemy, and a mission. The other had no map, no clear rules of engagement, and no clean extraction.

That second war followed him home. It crept into his relationships, his sense of identity, and his ability to regulate emotion and stress. It tested him in ways that no selection course ever could. This book lives in the space between those two wars. That is where Pete's story belongs.

Pete was born with salt water in his veins and a compass in his soul. Long before the Navy ever called his name, he belonged to the ocean. As a young man, he became a lifeguard in Huntington Beach—Surf City—where the Pacific was both a playground and proving ground. He performed rescues. He learned to read currents, weather, panic, and timing. Pete developed an ease in the water that cannot be taught. It can only be lived. That comfort would later save lives. It would also shape his role in Naval Special Warfare. Like many young men of his generation, Pete's civilian life ended abruptly when the draft arrived; his draft number was seventy-three, so he joined up instead of being drafted. He answered the call the way he answered everything: directly, without complaint, and on his own terms.

My own journey had begun earlier. By 1968, I had already spent a year in Vietnam as a lieutenant (JG), running Patrol Boat River operations with ten PBR crews engaged in daily combat. I understood risk, lethality, and consequences before I arrived in Coronado for Basic Underwater Demolition/SEAL training. I volunteered for the Teams because I wanted a higher standard—one where competence, discipline, and teamwork were non-negotiable. I passed the physical and mental screenings and was appointed BUD/S class officer for Class 47. We graduated in November of 1968. BUD/S was a turning point in my life, as it has been for every man who survives it. It strips you down to essentials. It removes the illusion. It forces you to confront who you are when comfort disappears. That is where I met Pete.

Pete became my swim coach. He took what he had learned as a lifeguard and translated it into survival. Open-ocean swimming is not about brute strength. It is about efficiency, rhythm, and mental calm under sustained stress. Pete understood water the way some men understand language. He taught me how to endure distance, conserve energy, and not panic when fatigue sets in.

We trained under men like legacy BUD/S instructors like Master Chief Terry "Mother" Moy, a physical powerhouse who set standards most of us struggled to meet. We lived in a cold so deep it altered cognition. Stage-one hypothermia turned lips blue and filled lungs with congestion. Stage two brought hallucinations. I remember seeing an elephant on the beach during a run—my brain misfiring under prolonged exposure. Stage three was fatal.

That environment reveals character. Everyone coped differently. Some shut down. Some lashed out. Some quit. Pete created. Most weekends, the men who still had energy disappeared into town. Pete stayed behind. I would walk into the barracks and find him wearing nothing but his bathing trunks, perched on a top bunk like a gargoyle, sketching relentlessly. He drew caricatures of instructors, of the class, of the misery itself. Those drawings became a pressure valve for all of us.

He captured Dick Pearson, our lead instructor and seasoned combat veteran, in that coiled-spring posture—just before he would pull a punch out of nowhere. He immortalized Vince Olivera—Black and Indian heritage, always moving, never idle—running full stride with a massive, hooked nose, a cigar clenched in his teeth, trailing smoke like a locomotive. He drew us as "sugar donuts," dripping wet and rolled in sand, lips blue, miserable, and ridiculous. The instructors were then shown Pete's drawings. Officially, they were offended. We were punished. Unofficially, they laughed—because Pete had captured something true. That was his gift. Even under maximum pressure, he translated chaos into meaning.

After BUD/S, Pete went to UDT-13. He deployed twice to Vietnam, serving six-month tours offshore, swimming into hostile shorelines at night to conduct hydrographic and beach reconnaissance for Marine landings. He worked as a cartographer, mapping terrain with waterproof slates, often under fire. UDT units were not trained for direct action the way SEAL platoons were—but that did not make the missions safe. Pete was shot at. He faced the same uncertainty and exposure as any man operating in contested environments.

In 1970, Pete served as a swim team leader during the recovery operation of the Apollo 13. During that mission, he hand-painted a cigar-chomping Frogman—"Freddy the Frog"—on recovery helicopter #66. That image traveled the world. It became iconic not because it was clever, but because it was authentic. It came from one of us.

Pete did not paint scenes. He captured the spirit of the Teams. Today, his work lives in places that understand its significance. The National Navy UDT-SEAL Museum in Fort Pierce, Florida, and the Naval Special Warfare Museum in Coronado, CA. The USS *Midway Museum* in San Diego house original Freddy the Frog artwork, large-scale murals, and historical illustrations created by Pete. Curators and historians recognize him as a combat artist—a man who preserved history visually, with accuracy and emotional truth. These murals are not decoration. They are records. They depict hardware, context, and lived experience. They tell the story of who we were when the Teams were still becoming themselves.

After retirement, Pete's life followed a path that was

anything but conventional. Our friendship continued through purpose, chaos, and loyalty. During that period, Pete came into possession of a 52-foot Chris-Craft that would become both shelter and symbol. The boat would eventually be known as *Patriot SEAL,* and the way Pete acquired it said everything about the way he lived.

The original owner had fallen behind. The boat had been anchored in San Diego Harbor for days before it would be tagged, towed, and hauled out in Chula Vista, where it sat on skids while port fees accumulated faster than anyone could pay them. The owner could not afford to retrieve the boat. What he could not afford to lose, however, was his welding equipment tied directly to his livelihood. Pete saw the way through.

He orchestrated a backdoor deal that reflected both his loyalty and his unconventional ethics. He paid the Port Authority fees and the dead-boat costs, returned the welding equipment to the owner, and quietly retained the boat. No formal sale in the traditional sense. No clean paperwork trail. Just an understanding built on necessity, trust, and timing. From that point on, the boat was always sailing under the radar—literally and figuratively.

Pete lived aboard *Patriot SEAL* at Fiddler's Cove. Maintenance was inconsistent. Compliance was optional. Independence was absolute. The boat mirrored the life Pete was living after the Teams: functional but frayed, resilient but nonconforming, held together by ingenuity rather than regulation. Eventually, the Harbor Police took notice. An APB circulated. Pete adapted the way he always had. He painted over the name. He altered the silhouette with planks. He made the vessel harder to identify, harder to categorize—harder to seize. When it became clear the boat would be taken from him, we reverted to instinct.

We slipped out of San Diego Harbor at night on one engine, without a compass. No moon. No radar. One of us slept while the other navigated by sound and feel. Interstate 5 hummed somewhere to the east. Camp Pendleton passed unseen as we drifted and adjusted, operating the way we always had—quietly, deliberately, and just outside the margins. We ran north toward Long Beach and, eventually, the boat made its way to Ensenada, Mexico. Pete kept it afloat with flotation bladders and powered it with solar-electric motors.

Over time, the metal and brass were each stripped away. What remained was a rogue vessel—patched together, unconventional, surviving on adaptation rather than permission. That boat was Pete. Always moving just below the surface. Always operational, even when fractured. Never fully anchored, but never entirely adrift. It reflected the second war he was fighting—the one without uniforms or formations—where survival required constant adjustment, vigilance, and creativity. It was not chaos. It was how Pete stayed afloat.

In later years, Pete became what I can only describe as a super patriot. He created murals displayed at the Pentagon, the White House, Naval installations in Virginia and Coronado, VFWs, and local landmarks. After September 11th, he worked relentlessly—often 24 to 36 hours straight airbrushing scenes that expressed national grief, resolve, and remembrance. He lived in campers and tents. He barely ate. He burned through compressors and airbrushes, discarding them when the work was finished. He created textures that looked like wood, steel, and battle-scarred metal. He was not an entrepreneur. He rarely had contracts.

Often, Pete was not paid for his art. He did not care. What mattered was that the story endured. Pete also contributed, alongside others, to the conceptual development of the Naval Special Warfare Insignia, the SEAL Trident. With his cartographic mind and artistic skill, he produced sketches and emblematic concepts during a period when the Teams needed a unifying symbol—one that represented mastery of sea, air, and land.

The anchor.

The Trident of Neptune.

The flintlock pistol.

The eagle.

That gold pin carries weight. It represents endurance, lethality, sacrifice, and belonging. Pete wore his Trident long after he retired. It was inseparable from his identity.

But Pete's life was not easy. He lived with bipolar swings—manic energy followed by deep depression. He had no shut-off valve. Stress affected him differently. He was unconventional, sometimes difficult, and often brilliant. He distrusted systems and avoided bureaucracy. He found comfort in dogs, music, familiar places, and creation. Art was not a hobby for Pete. It was a lifeline.

When trauma attacked all three legs of his foundation—mind, body, and spirit—he rebuilt them not through bravado or denial, but through creativity. Art allowed him to externalize what his mind could no longer organize. It turned memory into meaning instead of weight. That is why this book matters.

The Painted Frog is not a biography. It is not another war story. It is a human story about what happens when the missions end, and the real fight begins. It is about resilience, identity, and the quiet courage required to rebuild after invisible injuries. For many veterans, the war does not end when the shooting stops. It follows them home. It settles into the nervous system. It reshapes sleep, memory, impulse control, relationships, and identity. The greatest injury is often not the trauma itself, but the absence of a place where that trauma can be understood without explanation. Pete understood this.

Recovery does not begin with diagnosis alone. It begins with trust. With shared language. Being seen by people who do not require translation. Many veterans do not need to be told what is wrong with them. They already know something is broken. What they need is a path forward that respects who they were forged to be. *The Patriot SEAL Heritage Foundation* exists for that reason. It is grounded in the belief that PTSD and traumatic brain injuries are not personal failures, but operational consequences.

Invisible wounds earned in service. Healing happens in connection—in circles of trust, accountability, and shared experience. Pete believed that creativity, discipline, and service could rewire what trauma fractures. He lived that belief through his art and by showing up for his brothers when words failed.

For those who want recovery—not the erasure of the past, but its integration, the Foundation provides a pathway forward. One that honors the full arc of service. One that recognizes that rebuilding mind, body, and spirit is not a retreat. It is an advancement. Pete "The Pirate" Carolan lived without ego. He passed away with his Trident on his chest. Throughout his life, he gave us a face, a symbol, and a history. This book is part of the mission my friend left us with.

BASIC UNDERWATER DEMOLITION/SEAL TRAINING
CLASS 47
NAVPHIBSCOL BASIC UNDERWATER DEMOLITION SEAL TRAINING
NOVEMBER 1, 1968
SEAL TEAM ONE
50TH ANNIVERSARY REUNION
UNDERWATER DEMOLITION
Pete Carola
PATRIOT SEAL
16-19 AUGUST 2018 • CORONADO, CALIFORNIA
"Left my girl way out West — Told her a Frogman's life was best!"

Pete Carolan – Hand Sketch @1970 – Private Art Collection
Peggyann Morley @patriotsealheritagefoundation

A MESSAGE FROM THE PSHF PRESIDENT

MS. PEGGYANN MORLEY

President of Patriot SEAL , Heritage Foundation

Pete the Pirate, if you could only know how many times, I have sat with all your board members of the Foundation and how we all expected to hear your booming laugh or see you rounding the corner with some wild new idea. There is not a single day that goes by that I do not think about you, not just as the force and founder behind the charity you began, *Patriot SEAL Heritage Foundation,* my love, whose warmth made everything brighter. Everything I do here—every dollar we raise as a foundation, every combat military mural, artwork, and artifact we protect, every story we tell—is a direct reflection of the incredible light you brought into this world.

Pete was the kind of person who walked into a room

and instantly doubled its brightness, due to his sheer joy for life and the mischief that sparkled in his eyes. He always wore SEAL polo shirts, never without his Trident, a silent pledge of allegiance. This message, indeed, this whole book, is my tribute, as well as our written and combined works, to your memory and the astonishing life you led.

That life was rooted in the water. High school only amplified his natural water prowess; Pete Varsity Lettered in Water Polo all four years, a testament to his sheer dedication and skill that he carried right into Orange Coast College, where he also joined the rowing team. Those old yearbook photos of you truly captured the young, hungry athlete, and every summer found him exactly where he belonged: as a lifeguard at Tower #3 in Huntington Beach. He was a true creature of the ocean.

But then, those Olympic dreams of Pete's were abruptly replaced by the Navy SEAL Trident. The Vietnam War was raging, and when Pete drew a low draft card number, he did not hesitate; he enlisted in the Navy in April 1968 and proceeded from Basic to BUD/S training in July of 1968. He completed the transformation with the graduation of BUD/S Class 47 in November 1968: a water warrior had become a Navy SEAL, a Frogman of UDT- 13.

He served two arduous deployments in Vietnam with UDT-13 as the cartographer, conducting reconnaissance mapping and securing landing grounds for the Marines. Pete continued to serve with distinction through the 1970s, including time with the legendary SEAL Delivery Vehicle (SDV) teams, before his honorable discharge in 1973 and subsequent transfer to the Reserves. He served in the military for a total of 23 years.

Then, the shock of 9/11/2001 brought him back into active duty in 2001, finally retiring in 2007 at the age of sixty. Yet, Pete never truly left the community he loved. He became the SEAL community's artist, translating the powerful spirit of Naval Special Warfare into vibrant, unforgettable combat art. Watching him create was always a marvel and a gift.

His combat military murals and artwork are truly infamous and stand as a towering, colorful reminder of his legacy. They are everywhere: decorating SEAL reunions, commemorating heroes at U.S. Navy SEAL museums, and even holding places of honor in prestigious locations like the Pentagon and the White House during the Bush administration. Whenever I travel to Coronado and San Diego, CA, I see his handiwork, and it ensures that the stories of our heroes and his own commitment to them will live on in bright color for generations.

"Pete the Pirate," as he was known within his brotherhood, came to be during his BUD/S training days. Many of Pete's brothers have written chapters within this book to highlight the character and commitment of their friend and lifelong military brother.

Pete served as the lead swimmer of the NASA Apollo 13 capsule recovery team in 1970. He told me, *"I snuck down onto the ship's deck the night before we picked up the astronauts to leave a calling card from the SEAL swim team."* The now-infamous caricature of "Freddy the Frog" was painted on the side of the helicopter used to recover the astronauts aboard the USS *IWO JIMA.* The character was a morale booster and stood out as a national symbol for those who appreciate military history.

Pete and his team of Frogmen safely recovered the astronauts from the Pacific Ocean when the capsule touched down. The helicopter aircraft with "Freddy the Frog" is now on display at the *USS Midway Museum in San Diego,* CA. That dedication to preserving the SEAL legacy — the drive of both the warrior and the artist — is what led him to found *The Patriot SEAL Heritage Foundation.*

Towards the end of his life, even as he was dealing with debilitating health issues, Pete possessed an astonishing clarity of purpose. I watched him suffer, truly suffer, from the wounds that were not always visible: the things that happened to him in Vietnam and then the shock of 9/11. He struggled with Post-Traumatic Stress Disorder (PTSD) and periods of profound depression. It was heartbreaking to witness. It was out of this personal struggle that the Foundation's true inspiration was born. He realized that the most significant legacy was not just recording history but actively caring for the heroes who are still making it. He felt an intense call to help first responders and those veterans struggling with the invisible burdens of service.

The Patriot SEAL Heritage Foundation became his final, most important mission. That is why our board is composed of individuals who share their absolute passion and is made up of a core group committed to collaborating with veterans' organizations and other charities that provide tangible benefits and mental health programs. Every day, we honor his legacy not just by remembering the warrior but by fighting for the well-being of those who have sacrificed everything, just as my dear Pete did.

After those early years of high school that we shared and his long decades of service, Pete Carolan was a distant figure in my life. The Trident had taken him deep into a world few of us could ever truly touch, much less understand. I did not see much of him until after 2010, when our high school class began to reunite. He would suddenly appear at our lunches, events, and reunions, often for a brief time, and then disappear just as quickly, like a ghost of our past. It was always the same: people would inevitably spot that familiar, restless energy and that wide smile, and they would lean over to me, inevitably asking, "*Who was that?*" And the answer, delivered with a mix of awe, respect, and a little bit of mystery, was always: "*That's Pete Carolan, he's a Navy SEAL.*" A full explanation was not necessary; the title spoke volumes about the man.

Pete re-entered my life in a more significant way in April 2018. He showed up at a class luncheon, a familiar routine—always after the food had been served, never paying for the event! It was a running, affectionate joke that he had somehow *SEAL-ed* his way out of the check. After lunch, a group of us adjourned to a bar to continue the visit. Several hours later, only Pete, another classmate, and I were left, and we decided to move the conversation to Pete's favorite watering hole: Blackie's by the Sea. It was the quintessential dive bar, a place with history and grit, tucked right near the ocean. As I sat there, happily sandwiched between a Navy SEAL and a Green Beret, I certainly felt safe!

It was here, in this perfectly weathered spot, that I learned about another of Pete's enduring, quiet passions: he had done all the swag and the murals for the place. His legacy was painted on the walls. When Pete eventually passed, it was only fitting that they held a memorial for him right there—bagpipes and cheers, a proper toasting to a life well-lived in a place he considered his second home.

That evening was a turning point for us. Paul, our classmate, joined us back at my house to listen to records. Pete was a romantic at heart, a true contradiction to his warrior exterior, and a sincere lover of showtunes. *Camelot* was his favorite, and he would sing along to the CD movie soundtrack in his beautiful baritone voice. Hearing him sing "*If ever I should leave you*" still stirs my soul, a bittersweet prophecy of our future.

He just never left that night. We enjoyed each other's company, with a deeply shared history of childhood, and would often speak for hours about our time growing up. As a result, my spare bedroom quickly became "Pete's" for the weekend. At the time, he was living with a Sigma Chi college brother whose home was undergoing major construction, so the timing, as Pete's timing always seemed to be, was perfect. And so, the friendship and our partnership that would eventually bring about this Foundation were rekindled, not in a boardroom, but in the quiet, comforting glow of my living room, set to the tune of a musical.

Honor, for Pete, always included friendship. Pete was never married, and although we were both single, Pete never dated anyone else during the time we were together; he was very loyal. However, he did attend events, such as holidays and dinners, and other events solo.

It was not long after that sweet, musical evening that the seriousness of Pete's health slammed into our reality. I was in Orange County when Pete called me late in the afternoon. He was not well—faint, weak, and desperately needing a doctor. I immediately rushed him to Hoag Hospital in Newport Beach, CA. I could not believe what I was seeing; after fifty years, here was this vibrant man being treated for life-threatening conditions. When he asked me, with that classic Pete deflection, what I was reading on the clipboard, I had to be straight with him. "*You are really sick,*" I told him. He was battling sepsis, fighting against organ failure or worse.

Three days later, I brought him back home with me. It was clear he needed constant care and the steady, grounding presence of friends who felt like family. His previous living situation had been, quite literally, killing him. That hospital stay marked the real beginning of our life together.

For the next four years, we were a constant presence in each other's lives. He was hospitalized ten times that first year, fighting infection after infection, but he eventually stabilized in the safe, warm home I provided. We had a mature, comfortable arrangement. Our relationship was built on a fifty-year history of deep understanding and respect. We never stopped each other from living our best lives, independently and together, and our comfort zone was deeply established. We would joke about asking permission, knowing neither of us needed it, but it was a lighthearted confirmation of our mutual regard.

Pete filled the home with music, art, and surrounded both of us with friends and family at parties we would throw together for the holidays. He would construct pirate ships that glow on Halloween and put up Christmas lights and several trees to create a festive atmosphere. Life felt fun and made me feel as if there was always another adventure right around the corner for us both.

Pete, despite his health issues, was fundamentally a happy man, always running that extra mile for a friend or a cause. Yet, the darkness would sit with him, too. The PTSD would surface, bringing with it bursts of anger and the internal struggle I knew was tied to his heroism. He was a beautiful contradiction: a warrior who belted out showtunes, a man who saw combat but loved the romance of Camelot.

This duality was the key to his final, great purpose. He never seemed to paint for enjoyment; everything was a commission. He would take the assignment, sit with it for a couple of days to absorb the history and mission, and then work around the clock to finish it. It was pure magic to watch that fire ignite. His airbrush strokes were incredible; the colors and patterns brought his work alive on the canvas. Then, exhausted, he would sleep for two days straight. I always kept the coffee on for him while he painted around the clock; I was the silent partner to his relentless, passionate creation.

It was during these quiet hours, between his work and his struggles, that he voiced his profound realization. He came to believe that the intensity of his suffering—the things that happened to him in Vietnam and the heavy involvement with SEAL Teams after 9/11—

was also the source of the incredible creative genius in those world-renowned combat murals.

His pain was channeled into that art, giving it its brutal honesty and soul. He knew then that the most significant legacy was not just recording history but actively caring for the heroes who still walk among us. That intense call to help first responders and those veterans struggling with the invisible burdens of service has become the focus of *The Patriot SEAL Heritage Foundation*. Our mission, forged from his experience and dedicated to mental health programs, is his final, most enduring act of honor.

Soon after he moved in, Pete, still recovering from his first sepsis scare, agreed to deliver a painting to a Wounded Warrior event in Arizona. The event honored a veteran of the Hanoi Hilton POW camp. This connection was deep for him; Pete trained with the SEALS to go onto the USS GRAYBACK for the mission to rescue, however he did not go on the actual mission. SEAL Team ONE went in to rescue the Hanoi Hilton Prisoners of War. The painting that Pete created was his personal tribute to his brothers, an early glimpse into the power of the art he would leave behind.

I drove him to deliver the painting, and we stayed just over 24 hours. I knew he was not well. Upon our return, he was back in the hospital with a painful, swollen infection in his legs. This ongoing infection, which he suffered from about twice a year for 10-day stretches, was a painful reminder of his time in service, something many Vietnam Veterans continue to suffer from due to agent orange, a chemical that was sprayed to kill the jungle during the war.

I worked tirelessly with him to secure his 100% disability benefits. He was physically a broken man: a damaged heart (with three stents), bad lungs, a bad back, and those continuous leg infections. It was astounding how he maintained a whole, active life despite the chronic pain. He would complain, but whenever I suggested a doctor, he would famously say, *"I'll go tomorrow."*

Our life was filled with Pete's emotional highs and lows due to his PTSD. The outbursts and the anger were hard, and his brothers at the reunions would ask about them. Pete refused medication or counseling, believing it would diminish his art, dull his senses, and impede his creativity. His stubborn refusal to care for himself, while frustrating for me, stemmed from his fierce conviction that this rawness was essential to his creativity. While he was never violent with me, I knew he had been with others.

I tried to guide him toward mental wellness, but the topic was off-limits. He would leave in a dark rage, sometimes breaking things, like the night he threw his dinner into the sink and broke a plate after losing a bid on property. He came back days later and simply said, *"It's just my PTSD."* It was unacceptable, but I understood the depth of his internal war.

Despite the struggle, watching him create his combat murals and paintings was a gift for which I will always be grateful, having seen it firsthand. His one-of-a-kind military artwork is truly inspirational; his art was his authentic voice and his ultimate weapon against the internal war that raged within him. Pete was deeply connected to the patriotism that arose after the 9/11 attacks, primarily through his art: In 2011, Pete painted a mural titled *"America's Heroes of 9-11,"* cementing his role as an artist responding to national trauma.

Following 9/11, his art missions became "more patriotic." He was instrumental in painting and restoring military artifacts, such as refurbishing an iconic airplane at Lions Park in Costa Mesa to be "more historically correct" of its Korean War-era appearance. His pieces, such as the 8' by 6' murals of the *Alamo* and *Freedom,* are not just paintings; they are powerful documents of courage and sacrifice. They convey the sheer, raw emotional weight of the military experience, providing vital recognition and inspiration to countless veterans who often feel unseen and unheard.

His fraternity brother, Steve Mensinger, the former Mayor of Costa Mesa, CA, was always a steady presence, coming to Pete's rescue. Steve gave Pete the key to the city and honored him by presenting Pete to Sigma Chi as a Significant Sig. Steve employed Pete to paint murals in his senior living community such as the oversized mural he called, *"The Alamo,"* including incredible ocean scenes at Palm Island. Steve tried to put Pete on a 9-to-5 schedule, but the starving artist could not be constrained. He was a force of nature, driven by internal combustion and worked at his own pace and time.

Pete loved the brotherhood of the Los Caballeros Cowboys where his own equine therapy began in 2013 but was never prepared to go on the yearly ride in Catalina. Jim Cunningham was his sponsor and brought Pete in as his gunsel, much like Pete did for John Bennett when he sponsored him. At the last minute he would come to me and tell me how he needed help filing paperwork out for the trip application. I would call then President, Tom Crone, and he would walk me through the process for Pete.

Along with me paying the fee, filing the paperwork, and preparing everything he needed for the ride we all worked as a team to keep Pete on track. Those rides made him excited and happy, no more dark clouds or confusion and temper where he would lose his tracking. The horses helped to reset his nervous system and to calm his mind, body, and soul.

On the evening of August 6, 2022, his 75th birthday, we attended a black-tie event for a board member of the charity, April Egbert. She had a celebration of life for her mother and Pete wanted to attend the party to support her. He was incredibly weak, suffering from an internal bleed that we did not know about. I helped him dress that night, but the moment we arrived a light switch flipped and Pete came alive. "Pete the Pirate" was fully engaged, delighting the ladies at our table, including our host. He was in his glory, commanding the room with his presence. We agreed to attend the event only if he promised to see his doctor at once upon our return. He made the most of his birthday party, and we were among the last to leave. He was never more alive and handsome than this night.

Just two days later, on August 8, 2022, he was admitted to Hoag Hospital. He told everyone, up to his final day, that he would be home soon, and I honestly believed him. Pete fought hard in those last moments of his life. Across from his hospital bed was a painting that hung on the wall. A painting of the U.S. Navy SEALs UDT West Coast reunion poster, as seen on the foundation's official website.

His death was unforeseen and devastating to us all. The doctors found an intestinal bleed and removed a third of his colon to try to save his life. His body, ravaged by Agent Orange, could not fight another battle. His organs failed him, and Pete was now gone.

On August 27, 2022, we lost him. It was devastating. Pete had introduced me to many wonderful communities, including the SEALs, the Los Caballeros Cowboys, and his Sigma Chi Brothers, he had attended USC after he returned from the war in Vietnam. The Alumni from the high school he attended were front and center to be there to support me. When he passed, they embraced me completely.

I am now the President of *The Patriot SEAL Heritage Foundation,* which Pete founded and was accepted for in February 2022. He wanted to promote Navy SEALs, veterans, and first responders to create and heal. This purpose has since evolved, directly addressing the internal war that many of our warrior's face once they return home.

I am carrying the torch now, along with the board Pete created, by focusing the foundation's efforts on securing K9 training for veterans and providing charitable support to first responders, with a focus on mental health support and ongoing programs for them. Working with companies to create 3 Day Equine therapeutic retreats for Navy SEALs as a reset, this work ensures that Pete's spirit — the champion who always focused on helping others, not his own struggles — continues to fight for the mental well-being of the military community he held so dear.

Pete the Pirate. His legacy is his art. If Pete Carolan, the Frogman, the Combat Artist, were speaking now, his final words to his community would carry the weight of peace earned through immense struggle. Pete Carolan's most significant piece of art is not on a wall or a canvas; it is *The Patriot SEAL Heritage Foundation.* Now, the Foundation serves as the deployment strategy, enabling others to find their own canvas.

Pete may share with you, "*My art was never about glory. The canvas became the one place where I could translate the emotional weight of what we carried—the darkness, the sacrifice, the grief, and the unbreakable bond of brotherhood.*

Every mural, every brushstroke, holds memory. They are not decorations on a wall; they are reminders. Moments of silence are captured in color, so the cost of service is never forgotten in a world that moves too fast.

Look at the sea. The image of 'Freddy the Frog' I painted on the Apollo 13 recovery helicopter—now preserved aboard the USS Midway—may be one of the quietest statements I ever made. It carries humor, humility, and hope. It is reminder that even in moments of great danger, there is still room for light, for laughter, and for home.

That was the heart of a Frogman. Tough skin, soft heart. The mural at the National Navy UDT-SEAL Museum stands as a time capsule for every man who carried the spirit of service, sacrifice, and patriotism for this country. Not for fame. Not for recognition. But for the brotherhood, and for the generations who deserve to remember."

If I could speak to Pete one last time, I would say to always take better care of yourself because we are all missing you dearly. I can see signs of you everywhere inside of our high school classmates with conversation at our class reunions, the Los Caballeros gatherings, and charitable events we hold in your name within the foundation. We will continue to do excellent work in your name.

CONTENTS

BOARD OF THE PATRIOT SEAL HERITAGE FOUNDATION

PHILANTHROPIC PARTNERS

LOS CABALLEROS

AUTHOR'S PURPOSE

"To bridge the gap between history and the reader."

DR. (H.C.) LISA RAY

Ghostwriter / Publisher / Founder Golden Ray Publishing

I did not come to this storytelling by accident. Legacy stories arrive as opportunities—this one arrived as a responsibility. *The Painted Frog* was never meant to be simply a book about a man; it was meant to become a permanent record of a life lived in service, a brotherhood forged under pressure, and a legacy that has quietly shaped both military history and American culture without ever demanding recognition.

From the moment that I became entrusted with the story of Pete "the Pirate" Carolan by the *Patriot SEAL Heritage Foundation*, I understood that what lay ahead was not authorship—it was preservation. At its heart, *The Painted Frog* reveals the life of a man who refused to be defined by any one chapter of his journey and asks us all, what is the legacy you leave behind? Pete Carolan was a warrior, an artist, an adventurer, a brother, and a storyteller. He carried the spirit of the Frogman long after he took off the uniform.

This book reveals a man who endured loss, faced life's challenges with humor and grit, and found healing and purpose through creativity. His art became more than paintings—it became a bridge between generations, preserving the untold stories, traditions, and brotherhood of the UDT/SEAL community.

Pete's greatest legacy was not just the missions he completed or the murals he painted, but the way he lived: with curiosity, mischief, loyalty, humility, and a deep love for the men he called brothers.

In the end, *The Painted Frog* asks us to consider what we all leave behind. Pete's answer was simple: a life fully lived, stories worth telling, and a legacy that continues to inspire long after he is gone.

Collaborating with Retired Navy Commander and SEAL Team One Skip Crane, Chairman of the Board of *The Patriot SEAL Heritage Foundation*, has been both an honor and a privilege. His commitment to bringing together the brotherhood of those who knew Pete throughout his military career and beyond has been inspiring to me personally. My time on the ground with Skip and Carol Crane while researching and interviewing in Coronado, CA,—through private Special Warfare tours and conversations with members of BUD/S Class

42, 46, 47, & 49 as well as shared stories around the kitchen table—helped me understand the depth and richness of Pete's life, and that perspective is woven throughout these pages.

It was through the President of the foundation, Ms. PeggyAnn Morley, and the entire Board of Directors of *The Patriot SEAL Heritage Foundation* that I was given access not just to materials but to stewardship. This is not a passive board. This is a working body of individuals deeply committed to protecting the artwork and the legacy of the UDT and Navy SEAL community, while actively advancing programs that serve Navy SEALs, veterans, first responders, and their families, with foundational initiatives in mental health, TBI, and PTSD, with K9 training and horse therapy retreats and resources. Structured therapeutic programs for healing and hope, a return home to themselves.

The PSHF Board works with the same ethos that defines the men and women whose stories are told here: discipline, accountability, and mission-first execution. Their leadership ensures that this foundation is not simply reflective of the past but actively shapes the future. It is because of their vision and trust that this project was commissioned, supported, and brought to life with the level of integrity it demands. The open format conversations with board members: Andy McTigue, April Egbert Thompson, John Bennett, "Mickee" Hernandez, Carol Marshall Crane,

Kristopher Marshall, Deanna Green, Lance Green, and Tommy Bennett are reflected inside of the heartbeat of this book.

The partnerships with for-profit and non-profit establishments that pair those in need with trained dogs and healing with horses are the cornerstone of the foundation. A modern extension of that same brotherhood, carried forward by former Navy SEAL Team One, Andy McTigue, through the Veterans Canine Intelligence Academy (VCIA) 501(c)(3), where veterans train alongside their own dogs—reclaiming purpose, discipline, and the enduring mindset of service.

For profit partnerships that are committed to three-day retreats offering equine therapy with licensed therapists and horse-riding experts like Jeff & Monica Skelton in Catalina Island, CA, with the Catalina Island Horseback Adventures, ensure that funding for ongoing mental health rehabilitation initiatives will be met. Fast Passport Centers, whose shared commitment helps transform remembrance into action through programs benefiting veterans and first responders.

Among those connected to this mission is John Bennett, PSHF VP of Strategic Partnerships, whose path into this brotherhood reflects the same standard of excellence and commitment. Pete Carolan himself was a member of the Los Caballeros—a private and highly exclusive riding group whose membership is not sought out , but earned. It is not a club one joins; it is a distinction between a group which one is invited.

John Bennett studied directly under Pete to become a Los Caballeros—an achievement that speaks not only to skill, but to character, discipline, and trust. Pete's passing in 2022 made it so that John rode alone on his first outing. The men who supported Pete, now supporting John. That same alignment now carries into his role on the Board, further bridging Pete's personal legacy with the foundation's ongoing mission.

As I moved deeper into Pete's world, the process became something far more immersive than traditional research. It became an act of preservation. Every photograph, honor and challenge coin, swim medal, newspaper clipping, every piece of artwork, every document from four presidents spanning five generations of Presidents, International Space Station ALPHA—each one carefully digitized, cataloged, and protected and can be found in the closing remarks of *The Painted Frog*.

Hours turned into days, and days into months, as I built a living archive within Google Drive—a modern vault for a legacy that began long before the digital age. Hand drawings, sketches, and original paintings are now preserved to tour national libraries and military museums. What once lived in boxes, storage containers, in basements, in memory, now lives in permanence. And yet, the most powerful moments in this book did not come from files; they came from each contributor's voice who I interviewed to tell the unique story of Pete "the

Pirate" Carolan over the course of one year, 2025-2026.

Through direct interviews with every individual represented within these pages—men like Al Starr and (Ret.) Lieutenant Ernest Jahncke—I was given access to something no archive can hold: the lived experience of the Apollo 13 capsule recovery 1970 as told in the media on national news by Walter Cronkite. These were not rehearsed accounts. They were not curated for an audience. They were delivered the same way they were lived—directly, without embellishment, and without the need for recognition.

The men spoke candidly of Apollo 13 not as a headline, but as a mission. Of uncertainty. Executing under conditions with no margin for error, and for Al Starr, getting roped into another of Pete's adventures to airbrush the UDT 13 "*Freddy the Frog*" he created on the helicopter #66 that was used to lift the astronauts back to safety after splashdown. That same helicopter, now on permanent display at the USS Midway Museum in San Diego, CA. And in those conversations, the truth became undeniable: while the world celebrated the astronauts, it was the frogmen who completed their return.

Pete Carolan was one of those men.

He served our country with two tours in Vietnam as a UDT-13 cartographer, where precision mapping and reconnaissance meant the difference between mission success and failure, often under fire. He trained under great leaders like Vince Olivera and Master Chief Terry "Mother" Moy, both First Phase BUD/S instructors whose legacy is now permanently honored. Through my first published military offering; *MOY BELL—The Only Easy Day Was Yesterday*—the replica brass ship's bell, the "hell bell" of BUD/S hell week training and the iconic bell rung in remembrance at the West Coast Navy SEAL reunion for fallen SEALs yearly we learn ethos. This American symbol of the relentless standard needed to earn the right to stand among the Teams and wear the Trident, now on permanent display at the *San Diego Navy SEAL Museum.*

Pete carried that standard forward—not only in service, but in expression. Tom Murphy, author of "*Beyond the Trident,*" who gives the Opening Reflection of *The Painted Frog*, sent to me his cruise book of 1970 , Vietnam to reference in my investigation of "Pete" facts. Tom spoke to me about the 3-legged stool that supports all of us. Mind, Body, and Spirit. If the stool is not balanced, you will fall.

R.K. Bell tells us all of the human sacrifice of war and firsthand accounts of working with Pete in Vietnam as his Commanding Officer. Pete's drawing and sketches in the UDT 13 tour book tell the story of the tour. Personal, time-stamped iconic sketches and art that define the era.

When Pete was called back into the reserves his super patriot love for his country could be found in his art.

Pete’s large-scale 911 military murals found in San Diego, CA, depict the first responders who were lost in the 9/11 tragedy in New York. His nationally recognized military murals include the commissioned piece at the entrance of the world-renowned *National Navy UDT - SEAL Museum in Fort Pierce, FL.* Pete translated experience into imagery. His work does not simply decorate space—it defines it.

Pete, he was a super patriot through and through. Over Pete’s life he created several pieces of signed artwork that he gifted to his fellow BUD/S, UDT 13, Sigma Chi, Los Caballeros, and other super patriots’ friends in the Teams and brotherhood that he met along the way. His 9/11 memorial murals stands today as further testament to his unwavering patriotism for our first responders, capturing moments of national grief and resilience with the same authenticity he brought to every aspect of his life.

Throughout this project, I uncovered presidential letters of acknowledgement spanning decades—from Ronald Reagan to Donald J. Trump—each one reinforcing the national significance of Pete’s contributions. Pete’s artwork can be found at The Pentagon, the White House, Navy SEAL museums, VFW, and in and out of Naval War command stations, and in the favorite watering holes of our Special Forces where they go to reconnect to the community. Staples in Southern California like Mc. P’s in Coronado, CA, and Blackie’s in Newport Beach, CA, are amongst the most recognized. His works are scattered throughout our National Parks and grace the fire mantle of great warriors and mentors such as Rear Admiral / U.S. Navy SEAL, Ray Smith. The letters from each President, the commanders in chief, do not just confirm his story—they elevate it, placing him firmly within the continuum of American service and recognition.

Additional submissions by his BUD/S class 47 and Navy SEAL Team ONE/UDT 13 brothers, Charlie "Sandy Prouty, Tom "Rocky" Wagner and BUD/S class 49 Bill Pozzi, now under review by military institutions and museums, further connect Pete's hand-drawn sketches and the elements he drafted for review to the evolution and artistic interpretation of the military insignia now recognized as the Navy SEAL device, the Trident. This work to uncover his contributions is ongoing, and its significance continues to unfold, even today.

But the most important truth is this: Pete never did any of it for recognition. He created these pieces of military art to connect, and honestly, because it was who he was. He struggled with elements of mental health, PTSD, and traumatic brain injury. He fought the unseen battle our veterans may fight long after the war was over. This book exists to ensure that his story—and the stories of the men and women, his board of directors, and brotherhoods who stood beside him—are not lost.

The proceeds from *The Painted Frog* are donated directly to the *Patriot SEAL Heritage Foundation*, supporting the continuation of programs that serve Navy SEALs, veterans, first responders, preserve history, and extend the mission beyond the battlefield. This book is not the end of that mission—it is part of it.

Peter J. Carolan's legacy military artwork will tour alongside his book, bringing his visual legacy into the same spaces where his story is told. And in 2027, a companion tabletop art collection book special edition will be released further expanding the reach of his work and the depth of his impact.

I began this interview, ghostwriting, and book publishing process thinking I was preserving Pete's life through storytelling of those who knew him throughout his entire life. What I quickly came to understand is that I was being entrusted with something far more significant. I was being asked to carry forward a standard. A way of being. A level of responsibility that does not end when the mission is over.

There were moments in this process that were not easy over the last year for compiling and documenting the book project. Sitting across from men, now in their 80's, who spoke so plainly about extraordinary events, without embellishment, without drama—it forces you to confront your own belief of what matters. It strips away everything that is unnecessary and leaves you with something far more difficult to hold: Truth without performance. And in that truth, I found something I will carry for the rest of my life. That legacy is not loud. It is not self-declared. It is not dependent on recognition. It is built in the moments no one sees. It is carried forward by those willing to do the work without needing to be known for it, and that has become the true definition and meaning of integrity.

Pete Carolan lived that way. And now, having his story entrusted to me, I understand that my role does not end with the final page of this book. It begins there. Because preservation is not a single activity, it is a continuation. The archives we built, the interviews captured, the artwork restored, the history verified—these are not endpoints. They are foundations to support ongoing initiatives for PSHF.

There will be more stories in my future. More men. More legacies that require the same level of care, discipline, and truth in my storytelling. And I will approach them differently now. With even more respect and patience. With a deeper understanding of what it means to be entrusted with someone's military life legacy story. This work changed me. Not in a way that can be easily explained—but in a way that is now embedded in how I see the world, how I listen, how I write, and how I carry responsibility forward. I no longer see these stories as projects. I see them as assignments. And assignments are meant to be completed with precision, integrity, and follow-through.

That is what I learned as the Architect of "The Painted Frog." That is what has changed for me. And that is what I will carry forward as my lasting memory of Pete "the Pirate" Carolan, from the men of UDT-13, and from every single voice and contributor that trusted me enough to write with them, their spoken words.

Pete Carolan became a friend to me through the pages of this book. A man whom I will never meet. He came alive in the stories shared by those who rode beside him, served with him, laughed with him, painted with him, and welcomed me—without hesitation—into the world he helped create. Through countless interviews, conversations, trail rides, and Zoom calls, I came to understand that Pete's greatest masterpiece was never a painting. It was the life he built and the lives he touched along the way. His legacy now lives on through the *Patriot SEAL Heritage Foundation*, where his mission of brotherhood, service, healing, and purpose continues to inspire future generations.

While this book may reach its final page, Pete's story does not end here. The work continues. The friendships we have all formed continue. The impact inside of our national libraries and museums continues with art installations. If there is one lesson Pete leaves behind, it is this: live boldly, love deeply, create freely, and never stop becoming who you were meant to be.

Pete painted his life in full color. May you do the same. Because every day is a blank canvas, and each of us holds the brush. The world you create—through your choices, your character, your service, and your kindness—becomes the masterpiece you leave behind.

Paint it well, he did.

PROLOGUE

Pete "The Pirate" Carolan was one of those men who worked in the shadows, and he left behind a standard—the kind that cannot be framed, archived, or fully explained. It is the kind you step into quietly, without ceremony, and at once understand whether you belong to it or not. His name was never built for headlines. It lived where it mattered most—inside the Teams, inside the brotherhood, in the unspoken recognition between men who never needed language to confirm truth.

And yet, for all the silence around his name, there was always motion around him. Those who knew him remember something simple before anything legendary: Pete would sit before a mission or recovery—no speeches, no performance—just a quiet pause, like he was listening for something only he could hear. Then he would move. No ceremony. Only decision.

Pete Carolan's story begins in the saltwater discipline of Huntington Beach, where he served as a lifeguard forged in the same coastal grit that defined Southern

California's most elite watermen. Long before his name would be etched into history, he had already built the foundation of endurance, discipline, and rescue-level composure that lifeguarding demands. —an environment that demanded sharp instincts, physical resilience, and a quiet readiness for crisis long before he ever wore a uniform.

It was from that shoreline duty that he was drafted into military service, a turning point that moved him from civilian ocean rescue into the demands of UDT-13, where the same waterborne discipline was refined under the highest levels of Naval special operations training.

Before the world knew the term Navy SEAL, there were Frogmen—UDT 11| 12 | 13. No spotlight. No audience. Only work. Pete was forged there. He moved through water like it was memory rather than terrain, mapping hostile coastlines in Vietnam so others could come ashore safely, carrying the burden of visibility in a world that rewarded invisibility. Later, during the Apollo 13 capsule recovery, he was among those who entered the Pacific as men returned from the edge of space, part of the unseen effort that brought them home.

There was a cost to that life was never spoken of in real time. It was carried later—in distance, in silence, in the way men stop telling stories they once told freely. Pete did not escape that cost, and neither did those who stood beside him.

In a moment that never meant to be preserved, he painted a frog of UDT 13 on recovery helicopter #66. That helicopter now sits on permanent display at the USS Midway Museum in San Diego, CA. It was never meant to carry meaning beyond that instant, yet meaning does not always ask permission to endure. The painted frog, "Freddy the Frog" was Pete's way of honoring his brothers. A stamp with a deep meaning. *"We were here; we mattered."*

The national symbol of Frogman outlived the mission and settled quietly into history—not as decoration, but as witness. Pete Carolan's legacy does not live in operations or artifacts. It lives in men, his art, and now his foundation.

That same relentless ocean instinct carried him forward into collegiate competition at USC, where he joined Sigma Chi and paddled with the legendary Blazing Paddles crew—an era when California athletes began reshaping the world of outrigger canoe racing. In 1978, that team made history by becoming the first non-Hawaiian crew to win the 1978 Molokai-to-Oahu channel crossing, conquering one of the most punishing ocean races on earth and signaling a shift in the sport's global landscape. From lifeguard towers to collegiate paddling and finally into the Kaiwi Channel, Pete's journey culminated in a defining victory at Molokai—an achievement that placed him among the pioneers who proved that determination, teamwork, and ocean mastery could transcend geography and tradition.

Pete's military lives in those who served beside him and still speak his name with quiet certainty, as (Ret.) USN Commander Skip Crane reflects in the forward—"*not as myth, but as memory sharpened by time, where standards matter more than stories.*"

It is carried forward through the President of the Patriot SEAL Heritage Foundation, PeggyAnn Morley who ensures that what Pete represented is not preserved as remembrance alone but continued as responsibility creating partnerships and programs for Navy SEALs, Veterans, and First Responders. Those who struggle with the silent war of PTSD, TBI, and Related Conditions, just as Pete Carolan had.

The architect and publisher of the book further shapes it; *The Painted Frog,* Dr. (h.c.) Lisa Ray, who frames this not as biography, but as something still in motion and something that refuses to settle into the past because it continues to shape the present.

Tom Murphy, author of *"Beyond the Trident"* highlights the three-legged stool philosophy. If all three legs of your life, mental, physical, and emotional, are not addressed and cared for, your stool will collapse. This idea of nurturing ourselves highlighted in his contribution here in *The Painted Frog.*

Artist and creator of the Ronald Reagan *"Freedom Award"* Cheryl Allison captures what lingers between lives—the way one man's presence echoes across parallel paths long after contact ends.

Within the voices of (Ret.) USN CO, R.K. Bell and Class 47 BUD/S graduates, the record becomes a lived experience. Mike Sandlin speaks about the work itself, the formation of the device, the Trident, and what that work demanded. Ted Mosure carries the intersection of service and military artwork, where what is seen and what is remembered becomes inseparable. Charlie "Sandy" Prouty preserves Pete not as a symbol, but as a presence— "Pete the Pirate" as a living imprint of character. Tom "Rocky" Wagner speaks to a brotherhood that does not diminish with time, and Bill Pozzi remembers what was carried home when words were no longer enough, reminding us of all that the details in the TRIDENT were placed there by Pete Carolan. Gil "Espi" Espinoza UDTRA Class 42 shares the brotherhood beneath the surface and Jim Lavore BUD/S class 49 reminds all of us that Pete was a pirate at heart aboard The Patriot *SEAL*.

In the water, the story remains unbroken. Al "Smiley" Starr recalls Apollo 13 capsule recovery as responsibility rather than history, when recovery meant more than retrieval—it meant return. Lt. Ernest "Field Marshal" Jahncke carries forward what Frogmen have always known: trust is not declared; it is proven in silence.

Beyond the missions, others carry what Pete left into the years that followed. Andy McTigue, SEAL Team ONE speaks to being forged outside the lines, where identity is shaped by action rather than structure. Ann Kenney reveals the man behind the painted frog—not as legend, but as presence that became a symbol without intent. Steve Mensinger speaks to the tension between pirate and patriot, identities that coexisted without contradiction in Pete's nature.

PSHF board members speak to the legacy of Pete Carolan. Lance & Deanna Green remind us all how a family is chosen and the board members of PSHF became that chosen family for Pete. Board members April Egbert Thompson, John Bennett, Carol Marshall Crane, Kristopher Marshall, and Tommy Bennett give renditions of the "why" behind a board member of PSHF is so valuable for so many.

Jeff Skelton offers a quiet understanding that men are not bound by the era they come from, but continue to move through it—carried forward by purpose, memory, and the restorative power of equine therapy, which helps military heroes heal and find their way home. Tom Crone and Jim Cunningham reflect an unseen brotherhood through Los Caballeros, a private riding group on Catalina Island where Pete Carolan found peace of mind, connection, and a sense of grounding through the bonds of horsemanship and sacred land. Hunter Rusack, of the Wrigley family legacy, reminds us that the land we preserve is also the promise we pass forward, offering a fifth-generation perspective on stewardship, continuity, and the kind of legacy that is not inherited alone, but actively protected and carried into the future.

Together, these voices do not simply describe Pete "the Pirate" Carolan. They hold the shape of what he created around him. Because his legacy does not live in missions alone. It lives in what those missions' produced men who still carry him forward without needing to be reminded why, even as distance from them grows wider. That is where the cost stays. Not as loss alone, but as something unfinished, still interpreted by those who stay.

And that interpretation continues now through the foundation that Pete Carolan founded, the *Patriot SEAL Heritage Foundation,* where service is not remembered as an idea, but lived as responsibility. Where brotherhood is not nostalgia, but obligation. Where remembrance is not a conclusion, but continuation already in motion. This book is not a biography. It is a reckoning of impact.

A collection of voices tracing the outline of a man who never asked to be remembered but earned it anyway. And in doing so, he left something behind that did not end with him. Because in this community there is a truth that does not change with time: Legacy is not written. It is carried. And once carried, it does not stop moving forward.

OPENING REFLECTION

TOM MURPHY CAPT, USN (RET.) NAVAL SPECIAL WARFARE

Former SEAL Team Six Commander
A Conversation with Author of
"Beyond the Trident"

There are men whose presence fills a room long before they ever speak. Men forged not only by combat, command, and sacrifice, but by what comes afterward—the invisible wars no medal ever acknowledges. Their voices carry something different, not volume or ego, weight.

When Captain, USN (Ret.) Tom Murphy appeared on my Zoom screen; there was no need for introduction. I had just watched the three-hour interview with Jocko Willink on YouTube about the battles fought straight out of BUD/S training in Vietnam and then the climb to Commander of SEAL TEAM Six.

I learned from listening to both men speak of the price and sacrifice of special operations training, and the toll it takes on its warriors as they reviewed Tom Murphy’s book, *Beyond the Trident.* His service record already spoke for decades in Naval Special Warfare, leadership at the highest levels, combat, command, brotherhood, and the silent burdens that often go with all three. But that was not why I wanted this conversation.

I wanted to understand what happens after a warrior takes off the uniform and fights the mental battles of PTSD | TBI | and related causes. COB of PSHF, Skip Crane emailed me a link to Tom Murphy’s book: *Beyond the Trident* along with his personal number and email and wrote to me, *“Lisa, Speak to Tom, he adds a layer to this book no one else can.”* Before I spoke on the phone with Mr. Murphy, I set out to learn all I could about his experiences, journey, motivation toward healing PTSD | TBI | and related conditions, firsthand.

As a ghostwriter, I have learned that the stories that matter most are rarely found in official biographies. They live in the pauses between words. In the memories men do not volunteer. In the moments where rank disappears and only truth remains. Like all great men from our Vietnam War generation, Tom was humble, approachable, unfiltered, and filled with knowledge that spans five generations of military experience. He is irreplaceable. A man amongst leaders who was forged inside of the TEAMS and has lived, explored, and now continues to journey to places that are healing, mindful, and primarily, truthful.

And like every great warrior story, it began with another man. Pete "the Pirate" Carolan. Tom Murphy spoke of him not as a memory confined to the past, but as a presence still actively shaping the men who had once served beside him. In Pete's sketches, in his humor, in the way he mapped chaos into something human and enduring, Tom saw more than artistry. He saw documentation of a world that no official record could fully capture.

Then Tom reached back even further. Onto his personal bookshelf in his office during our call. He produced the cruise book of Pete Carolan as a UDT 13 Cartographer from his tours in Vietnam. I had heard of this cruise book from several of his BUD/S Class 47 brothers, no one knew of its whereabouts. I could hardly believe my eyes when Tom turned to the camera proudly showing me the frogs of Pete's. *"Hello, my pirate friend, you are still with us,"* I thought at that very moment.

This official cruise book Pete created during the 1970 Vietnam tour as a cartographer—pages that now exist as something far greater than paper and ink. Tom described it as a living, breathing time capsule. Not simply a collection of images or notes, but a layered record of a man translating war, environment, and brotherhood into visual memory. Maps. Illustrations. Margins filled with humor, observation, and the unmistakable hand of someone trying to preserve meaning inside the uncertainty of service.

In Tom's words, it was clear: *"This book was not nostalgia. It was preservation. A record of presence. A record of a man who understood, even then, that time would move faster than memory unless someone gave it form."*

That cruise book, he explained, still holds the depth of Pete's mind at work—equal parts technical skill, creative instinct, and the quiet emotional intelligence of someone who saw his brothers not just as operators, but as human beings moving through a moment in history that would one day need to be understood, not just remembered. And now, decades later, that same artifact is being entrusted forward.

Tom Murphy is sending it to be archived within the *Patriot SEAL Heritage Foundation*—ensuring it is not lost to time but preserved as part of a living record of the men who shaped, endured, and defined that era of Naval Special Warfare history.

As he spoke, it became clear that this was no longer just about Pete Carolan, or even about one cruise book. It was about what survives when a generation of warriors begins to pass their stories forward, not just in words but in artifacts. In sketches and in maps drawn by hand. In books that still carry the fingerprints of the men who created them.

And in that act of preservation, Tom made something unmistakably clear: That legacy is not what is left behind when a man is gone. It is what is recognized, protected, and carried forward by those who understand its value while it is still in their hands.

Because once a story like Pete's is properly seen—not as decoration, but as documentation stops belonging to the past. It becomes part of something still unfolding. And for Tom Murphy, and for those entrusted with its stewardship, that responsibility is no longer abstract. It is personal. And it is permanent.

Below is an excerpt from our first Zoom, one of what became several calls. From there, the conversation resumed in real time.

MAY 16, 2026, ZOOM CALL
DR. (H.C.) LISA RAY & TOM MURPHY

LR: Tom, one of the reasons we are here today is because of Pete 'the Pirate" Carolan. Since his passing in 2022, I have had the privilege of collaborating with the *Patriot SEAL Heritage Foundation* over the last year to ghost-write, interview, and publish his book *The Painted Frog* and to preserve not only his artwork, but the legacy of the man himself. The foundation has commissioned me to manage the architecture, preservation, and the writing of *The Painted Frog* to ensure future generations understand what Pete contributed—not just as an artist, but as a Frogman and storyteller.

Q1: You mentioned you still have Pete's original UDT-13 cruise book, filled with his artwork. What does it mean to hold onto something like that after all these years?

TM: *Well, Pete did 85% the artwork in that cruise book. Every time I open it, I see him. Not just his drawings—but him. His humor. His creativity. His personality. And when* *you have carried something like that for decades, it is not just a book anymore. It becomes a piece of your life. A piece of your brothers. A piece of your history. When I heard what you were doing—preserving Pete's story, scanning his artwork, telling the larger story—I thought, this belongs with you. Because this is not nostalgic. This is a legacy. And legacy matters.*

LR: That is exactly why this project matters to me. Pete did not just paint walls—he painted emotion. Through cartoons, murals, and military humor, he found a way to express things most warriors never talk about.

Q2: I know after your own retirement, you began painting as well. As someone who has spent a lifetime in service, what did creativity unlock for you?

TM: *I think anything that brings out the creative side of a person is important. After I retired—retired for the third time—I started painting. And I got enormous benefit from it. It opened something in me. Something that had been sitting there for a long time. I think for a lot of warriors, we spend so much of our lives working, moving, planning, and executing that we do not always make room for that creative side. But creativity reaches places that talking sometimes cannot. Painting. Writing. Music. Any of it. It matters. Especially for people who have trauma.*

LR: That resonates deeply with what we are building through the *Patriot SEAL Heritage Foundation.* We have been commissioned to expand beyond traditional support models into art, music, and now other healing modalities for veterans with PTSD and TBI.

Q3: And what I'm hearing from you is that creativity isn't a hobby, It's medicine.

TM: *Exactly. It is another doorway. And for some people, it is the doorway.*

LR: In *Beyond the Trident,* you talk about what you call the three-legged stool—mind, body, and spirit. That concept really struck me.

Q4: Can you explain where that came from?

TM: *That came from my own life. Most warriors have the body managed. Most warriors have the mind managed. We are trained for that. We are conditioned for that. But the spiritual side? That is often dormant. Or ignored. Or atrophied.*

And the truth is—if one leg of the stool is missing, eventually the stool falls over. That is what I discovered. Growing up, all three parts of me were active. I was physically active. Mentally strong. Spiritually grounded.

But by the time I became a SEAL, the spiritual side

went to sleep. Not because I stopped believing in something bigger. But because combat does not leave room for that. The mission comes first. The brotherhood comes first. Survival comes first.

And years later when the uniform comes off... That is when the questions start showing up. Who am I now? What parts of myself did I leave behind? What do I do with all of this?

That is when I realized my spiritual leg needed to wake back up. And honestly... That leg ended up saving me.

LR: That is powerful—especially in the SEAL community, where so much identity is built around performance, resilience, and mission.

Q5: What you are saying is strength alone does not heal trauma. Balance does?

TM: *Exactly, Real strength is balance.*

LR: Through my work as a ghostwriter, I have interviewed warriors from multiple generations, and one thing I hear repeatedly is this idea that the real war begins when the deployment ends.

Q6: What do people misunderstand about PTSD—especially within the SEAL community?

TM: *They think it is a weakness, it is not. PTSD is what happens when the human nervous system carries experiences it was not designed to process alone. And in the Teams, brotherhood keeps you alive, that bond becomes everything. You trust those guys with your life. You would die for them. But when you leave... You do not just lose a career. You lose your tribe. You lose your mission. You lose your identity. And that's when things can get dangerous. Toxic coping methods such as.*

Alcohol.
Isolation.
Anger.
Addiction.
Depression.

I have seen all of it. Sometimes it shows up right away. Sometimes it takes twenty years. But eventually if you do not address it, you can guarantee, it finds you.

Q7: And trauma itself became the catalyst that led you toward spirituality, consciousness, and a much deeper understanding of healing?

TM: *Absolutely. Trauma opened the door. And curiosity kept me walking through it."*

LR: You have spent years now studying spiritual teachings, consciousness, and what it means to truly heal. That is not a path many people would expect from a former Navy SEAL captain.

Q8: What did you discover?

TM: *I started reading everything: Jesus. Buddha. Krishna. Moses. St. Francis. Different religions. Diverse cultures. Different languages. And what I found was remarkable.*

> *Same message:*
> *Compassion.*
> *Service.*
> *Brotherhood.*
> *Forgiveness.*
> *Love.*

The fundamentals were universal. That is what eventually led me to working on what I call a spiritual manual. Because healing does not happen in one breakthrough. Healing is practice. It is how you live. How you serve. How you stay connected. How do you ask for help? How you keep growing. That is what matters.

Q9: So for the warrior sitting at home right now—dealing with PTSD, TBI, addiction, or feeling completely lost—what would you tell him?

TM: *I would tell him this, "Keep moving. Stay curious. Try something healthy. Paint. Write. Meditate. Read. Serve somebody. But do not isolate yourself. And do not ignore the spiritual side of yourself. Because that part of you…*

Has been waiting for you all along.

4 Horsemen of the Apocalypse

PARALLEL LIVES

CHERYL ALLISON

Ronald Reagan Freedom Award Design Artist

Peter Joseph Carolan was a U.S. Navy SEAL, Vietnam veteran, gifted artist, outstanding athlete—and an enigma. The story of his extraordinary life did not end with his death or the number of celebrated artistic projects he left behind. Instead, Peter's story will continue, starting with eye-witness accounts of events that seriously conflicted his life, and how he used his artistic talent to survive episodes of debilitating stress disorders and manic aggressive behavior that progressively became a constant challenge to control.

We never understood why our lives so often intersected in ways that were far beyond mere coincidence. Our respective families moved independently from the East Coast of the U.S. to Los Angeles County, California. A few years later, both families moved again, from Los Angeles County to Magnolia Street in Orange County, where my story about Peter begins. The unknown force that finally led us to live on the same street was more than geographic symmetry.

Magnolia Street is where Peter and I shared many personal, artistic, and familial conditions: absent, career-driven fathers, devoted mothers, unstable homes, and mature dedication to developing artistic talent. Our fathers were absent from home in ways that affect children most profoundly. Peter's father was in the Navy, and was often away at sea on long deployments. But upon returning home his presence was often the cause of tension, rather than comfort. My family details were different, but similar in impact. My father's career was centered in America's fledgling aerospace program, when missiles exploded on the launch pad and guidance systems routinely failed, lives were being lost in out exploration of space.

To correct these issues, certain engineers were often given short notice they were being moved to a new and sometimes distant location for research or to confirm vendor quality control. So, while Peter's home was affected by episodes of strife, parental arguments, and a horrific tragedy, my family moved around like Bedouin nomads, which caused my sister and I to attend fourteen different schools, adapt quickly and make new friends.

The year I lived on Magnolia Street is where I understood the Carolan family, and recognized something distinct about Peter. There was an intensity in his focus, a clarity in his vision. I remember sensing—not consciously, but instinctively—that he was meant to leave something important behind; that his life would extend beyond itself in some lasting way. Today, I still have pieces of his art, sketches from those early years—drawings, schoolwork, and later—military items, all carefully stored, and I will now be donating all these items to the *Patriot SEAL Heritage Foundation* for preservation.

Rita, his mother, was the emotional center of everything in his home. She was warm, artistic, and devoted to her children, especially Peter, the oldest of her three boys. Before Rita moved to California, Rita and her sister were inseparable, and billed as "The Coy Sisters," they performed tap dancing routines at places like county fairs, nursing homes, and charity events. The family's talent for creating art and dance routines directly influenced Peter's ability as an artist and speaker. They even contributed Disney character design. Art was not separate from life in that home; it was part of its foundation.

Rita took great pride in her homemaking skill. After moving to Magnolia Street, and in addition to their own backyard pool, Rita provided her three boys with swimming and diving lessons at Dr. Sammy Lee's Swim School in Anaheim. Dr. Lee won Olympic gold medals in swimming and diving (London, 1948), and was also a respected medical doctor who retired from practice in 1990. Dr. Lee was Peter's first professional swim & dive trainer, and how at an early age, Peter began building the discipline and endurance that would later define him in the waters of Vietnam, UDT 13 and Apollo 13 Capsule Recovery.

Junior Olympics 1960. Peter Carolan top row 3rd from left.

But beneath the creativity of that home, there also existed the pressure of his father's emotional volatility—when tears and shouting were punctuated by doors slamming shut, flying objects, and broken dishes. Where Rita nurtured, his father erupted, and physically focused his anger upon Peter—while Peter's two younger brothers rarely suffered from their father's wrath. Rita tried to compensate for family discord, soften what she could not control, but the constant emotional and physical damage Peter endured did not dissipate. It affected him deeply.

The few halcyon days of Peter's youth, and Rita's dependence upon her sister for companionship and emotional support, suddenly came to an end; the result of a tragedy from which the family would never recover. Rita's sister married a man who became a police officer, and from the outside, the marriage appeared stable. But inside that home, there were fractures—episodes of anger that mirrored what Pete already knew too well.

One night, her husband came home with his police issued service revolver, and what had begun as an argument escalated irreversibly to the point of no return. In a fit of rage, Rita's sister used her husband's police revolver to shoot him to death, and then she took her own life.

That moment did not end with the sound of gunfire. It expanded outward; grief and loss of her sister consumed Rita in a way that never fully released its grip, and the aftermath—funerals, legal matters, disposing of personal property, financial strain—it dragged on, prolonging the trauma.

The local story went to National headlines and further erased the family's fragile unity. For Pete, those years between ten and fifteen became defined by loss and instability. His father Tom eventually withdrew under the weight of it all, and the marriage of his own parents ended in divorce. What remained was a boy trying to make sense of a world that became intolerable.

Rita tried to redirect Peter, to give him something steady. She enrolled him in Boy Scouts, hoping the structure would help. Instead, it became the site of another life-altering event for Peter. While camping in the mountains, Peter fell into a deep ravine. The injuries were severe, and after a long search, Pete found unconscious and alone.

Years later, Rita told me that when Peter fell down the ravine, that was the point at which his demeanor and emotional stability changed dramatically. Rita said she believed Peter had suffered some kind of traumatic brain injury. Medical technology and patient care at the time did what it could, but it could not diagnose manic behavior or why Peter's behavior had fundamental changed.

Before the fall, Pete had been calm, thoughtful, composed. Afterward, his ability to regulate his emotions shifted dramatically. Outbursts came without warning, disproportionate to the situation. It was not simply behavior—it was as if something inside him had become disrupted. Today, we could possibly recognize it as a traumatic brain injury, TBI, or PTSD, but at the time, his manic behavior went undiagnosed.

The change in Peter was devastating to watch. He would create something beautiful and then suddenly destroy it, tearing apart his own artwork as if trying to silence something within himself. Rita sought help, taking him to doctors who prescribed Ritalin and other medications, but nothing seemed to stabilize him.

Without a proper diagnosis, every attempt was ineffective. The one place he found some control was in athletics. He excelled as a natural athlete with extreme focus. In the water, especially, he could focus, in the silence would find there. The discipline of swimming in their backyard pool and training by Dr. Sammy Lee—gave him a structure he could rely upon when everything else felt unpredictable.

As we moved through our high school years, our lives began to diverge. Peter and I attended different high schools, and we became more distant, staying loosely connected through our mutual friends. As the years rolled on, from time to time, we would reconnect, and he would tell me about Hawaii, about winning the Molokai race, about pushing himself physically in ways that seemed to give him balance. There was pride in those moments, but also a sense that he was searching for stability through achievement.

It was during our time connected to Orange Coast College that another piece of his identity took shape. Pete and I both attended there, still bonded through art, still moving in and out of each other's lives. It was there that he earned the nickname "The Pirate." One night, in a moment that captured both his boldness and his unpredictability, Pete climbed the school's flagpole and captured the pirate flag. When he was drafted into the military, he stuffed that Pirate Flag into his bag and brought a piece of home with him. Pete 'the Pirate' Carolan is a name that followed him into the SEAL Teams, becoming part of the identity others would come to know.

Then came Apollo 13 Capsule Recovery, and suddenly Pete was no longer just someone I had known, he was someone the community at large recognized. His name appeared in newspapers, tied to a moment of national significance. I clipped every article I could find, holding onto them not because of the headlines, but because I knew the person behind them.

Years later, in the 1990s, our lives converged again in a more significant way. We moved in together in Anaheim, carrying with us the history of everything we had shared. We had familiarity and connection, but there was also a reality I could not ignore. Pete's emotional responses remained unpredictable, and his jealousy surfaced quickly, often escalating without warning.

One night, a call came from a male friend I knew through an art class. It was a brief, harmless, and ordinary conversation. But when I hung up, something in Pete erupted, just like his father was described, breaking things in anger. Without a word, he began opening kitchen cabinets and smashing everything inside, firing one after another like grenades until nothing remained intact. I stood still silently waiting for the storm to pass. When it did, he seemed disoriented, as if trying to understand what had just happened. Then came the apologies, genuine and heartfelt. But sincerity did not prevent recurrences.

There were moments after these episodes when he would collapse completely, lying on the floor, motionless. It was as if he disappeared into the ground itself. I would sit beside him, bring a wet cloth, cover him with a blanket, and wait. Peter would not move for hours. When he returned to normal, he often had no memory of what had happened.

It felt less like emotion and more like something neurological— beyond his control. My visiting friends became concerned when I tried to explain we only use paper plates and cups, and that Peter prefers natural air conditioning by creating big holes in our walls. I will not even attempt to write about his ideas for interior decorating.

The final moment came at Newport harbor, after he had moved onto his boat, *The Senator.* By then, he had become increasingly isolated. I heard reports of his arrests, his fights at various bars, and that several bouncers were always required to subdue Peter before they could escort him to their executive office—otherwise known as "the street!" Peter also had a collection of photographic souvenirs from these violent confrontations, featuring his own two black eyes, bruises, stitches, bandages, arm slings; plaster casts, crutches, and wheelchairs—but nothing prevented future episodes.

I agreed to meet him at the marina where his boat was docked, hoping to help him get resources through the veterans, to reach him in a way that might help him. At that time, I was a police officer, carrying my service weapon in my purse. As I walked down the dock toward the boat, he suddenly leapt from the boat onto the dock where he charged at me without provocation. Before I could react, he picked me up and threw me off of the dock and into the water of the marina. The humiliation was immediate and complete as many were sitting having lunch, listening to music, and observed my involuntary baptism.

People inside the restaurant saw it happen and rushed to help me climb out of the water with the use of the ladder from the dock. I was soaked and shaken. Pete turned around and without a word, walked back to his boat while I walked up to the restaurant, dripping wet, fully aware of every eye on me, and called my sister to come and pick me up.

I never reported the incident, not to anyone. At that moment, I understood that Peter needed help beyond anything I could ever provide, and that his inability to control violent and dangerous behavior was now extended to me. I made the decision that I had to walk away.

A few years later, I met Navy SEAL Commander Skip Crane at the VFW in Coronado, and learned that he had been Peter's commanding officer. While speaking with Skip, I began to understand something I had long sensed. Peter respected him in a way that was grounded and had a steady kind of respect that reflected the father figure Peter had always needed.

Skip provided structure, accountability, and a sense of stability that gave Peter moments of clarity, a safe place within the chaos. Skip's authority and intervention literally saved Peter's life. He created boundaries, had him work with veteran services, obtained resources for shelter, food, medical attention, and started the process of therapeutic counseling, to restore Peter's mental health.

I saw that version of clarity years earlier when I attended an event with Peter in Huntington Beach, California. Peter stepped onto a stage, without invitation, and began to speak. Celebrities and astronauts had already given their speech and had exited the stage. Peter seized the opportunity to walk out on the stage, grab the mic, and command the room.

He introduced himself as a Navy SEAL and began talking about BUD/S training, about Hell Week, about Apollo 13, weaving humor, and intensity together in a way that captivated the crowd. Peter spoke of midnight maneuvers inside of training and successful missions as a UDT 13 cartographer in Vietnam. He described the kind of practical jokes and tricks SEAL's do to each other for fun, which police officers recognize as a felony.

The audience was leaving the event, but spontaneously returned to their seats, drawn to him, responding to his presence and his voice. At that moment, they saw and heard a hero. And he was. But what they could not see was the full story behind that image.

My own life continued to evolve through art, eventually leading me to study under Nick Pasko, a highly regarded Laguna Beach artist known for his disciplined approach and selective mentorship. He accepted only five students at a time and, until that point, had never taken on a female student. Yet he made an exception after my father took a ride to his studio to show him my work.

Under his guidance, I refined my craft and went on to become an accomplished artist, creating work that would reach national recognition, including contributions connected to former President Ronald Reagan. Our lives intersected again in an extraordinary way through that work. I designed the *Freedom Award, the highest distinction given by the Ronald Reagan Presidential Foundation* to individuals who have made monumental and lasting contributions to freedom worldwide. President Reagan himself was the first recipient, and the award reflected his enduring belief that one individual can change the course of history. That belief—so central to Reagan's legacy—was something I had witnessed firsthand in Pete, long before the world recognized it.

Pete's legacy lives not only in his service, but in his art. Over five decades, he received letters of recognition and gratitude from presidents across generations—from Ronald Reagan to Donald Trump—each acknowledging and thanking him for the impact of what he created. His artwork, born from both brilliance and struggle, will continue to live on, commissioned and displayed in national museums as a testament to a life that refused to be defined by any single moment,

What is often unseen, however, is the cost carried behind that legacy. When I sat down to share this story with Dr. (h.c.) Lisa Ray, the weight of it had not lessened with time. The memories remained vivid, the emotions close to the surface. There are stories that do not fade—they wait. And when they are finally spoken, they carry not just truth, but release. This is one of those stories and times in my life.

Pete Carolan was more than the man the world saw. He was a child shaped by trauma, a young man altered by injury, and a warrior who carried battles long after the mission ended. His life reflects something much larger—the unseen war fought every day by veterans and first responders living with PTSD, traumatic brain injuries, and the lasting imprint of military service. That is why Peter's story must do more than remember, it must act.

The mission now belongs to all of us. Pete founded the *Patriot SEAL Heritage Foundation* to meet that responsibility—providing equine therapy retreats and K9 training programs for Navy SEALs, veterans, and first responders who are still fighting the unseen war of trauma. These programs are not luxuries. They are lifelines. They offer structure, connection, and healing where traditional systems have too often fallen short.

We cannot continue to lose those who have already given so much. We cannot ignore the rising tide of suicide among our military and first responder communities, and we cannot expect them to carry this alone. Pete's story is not just a reflection of the past, but rather it is a call to the present. A call to recognize, to support, and to stand in the gap where healing must happen. His life, his art, and his struggle remind us that behind every act of service is a human being who deserves to be seen, understood, and supported long after the uniform is gone.

If his story moves you, let it move you to action. Support the mission. Fund the programs. Stand with those who have stood for us. Because the greatest honor we can give men likes Pete "the Pirate" Carolan is not just to remember them, but to ensure that others like him are never left to fight alone.

THE PAINTED FROG

POET - HEATHER HERNANDEZ

Forget how to quit
That was the training
Pounded into his very essence
There was something undeniably Pete
Whether saving lives as a lifeguard or a
SEAL in formation

The capsule splashes down
Nations breathe relieved
Apollo's return expose canvas
Pete's Dynamism affirms
Freddy's permanent aluminum form

The roar of the waves dilutes severity ashore
Strengthened through brotherhood reveals legacy
Advancing aptitude, his allegory
The sea cradled him with sweet reprieve

Whispering gently, a safe place to settle
Changing the Chemistry
Turning loud pain to beauty
Painting a dark shadow
Erupting deep inside
The voices of men lost

6 months of BUD/S
Basic Underwater Demolition
SEALS forming bonds

where men are stripped down
to what cannot break,
and rebuilt in cold silence
and shared suffering.

Brotherhood became breath.
Endurance became language.
And quitting ceased to exist.

What was once held in silence
found its way into color.
What was forged in pressure
learned how to speak.

The murals that graced Pentagon halls
And White House walls
calling to the masses of

The Unsinkable Spirit
That speaks of the desire to keep creating
And now it reaches me—
not as memory,
but as inheritance.
A pencil in my hand,
not given, but passed through time.
Weight without words.
Instruction without explanation.
The quiet command beneath every line:

So, I write the way he moved through water—
with discipline,
with surrender only to purpose,
with the understanding that creation
is its own kind of survival.

And I keep going because he did

BROTHERHOOD ACROSS WATER, AIR, AND SPACE

CAPTAIN R.K. BELL, USN (RET.)

I am Captain Ronald K. Bell, United States Navy, (Ret.). I graduated from Naval Special Warfare training with Class 25 at Little Creek, Virginia, in April of 1961. My career carried me through multiple assignments in Naval Special Warfare, including intelligence training at the Defense Intelligence School in Washington, D.C., and operational service as Assistant to the OP-34 Commander in MACV-SOG in Vietnam.

In May of 1970, I assumed command of Underwater Demolition Team Thirteen. In 1971 I commanded Underwater Demolition Team 11. In 1978 I assumed command of Naval Special Warfare Unit ONE in Subic Bay, PI. In 1980, after promotion to CAPT I commanded Special Warfare Group II in Little Creek, VA.

I have spent a lifetime in the Teams. When I reflect on those years now, what remains most vivid is not the machinery of war or the evolution of tactics, but the men. Their discipline, their humor, their resilience under pressure, and the quiet professionalism with which they carried themselves in circumstances most people will never encounter. Vietnam remains particularly present in my memory. Time has not softened it.

One loss remains especially close. Billy Machen served under my command as Officer in Charge of Detachment Golf, in 1966 during operations in the Nha Be area. He was a Platoon point man—first forward, first exposed, and first to make contact when contact came. In a complex ambush, Billy performed exactly as trained.

He triggered the enemy position, allowing the remainder of the team to react and survive the initial burst of fire. Roger Moscone moved forward through enemy fire to reach him. Billy was brought back alive, but he died during extraction. He was the first SEAL killed in combat. I was later required to identify him. Just a few short days later a telegram arrived addressed to both Billy and me . Our wives both delivered our first babies. Billy had a son, my wife Felicia and I, a daughter. It would be 50 years or so later that I would meet Billy's son at the West Coast Navy SEAL reunion. A full circle moment that has stayed with me a lifetime.

This operation, Billy dying in combat operations, was among the earliest initiated by SEAL personnel during a transitional period when Underwater Demolition

Teams were evolving into what would become the SEAL Teams. Those moments never leave a commanding officer. They become part of how you understand responsibility itself.

Leadership in combat is never abstract. It is personal, immediate, and permanent in its consequences. Over time, I came to understand that survival in our community is not only physical. It is also a memory. Leadership is what is preserved when operations end and when the noise of war fades. That understanding shaped how I viewed the importance of history, documentation, and ultimately, legacy. Pete "the Pirate" Carolan understood this better than most.

Pete was a SEAL of UDT-13, but he was also an artist in the truest sense of the word. He possessed an uncommon ability to capture not just events, but essence. During his time with the unit, he created a cruise book filled with original hand-drawn illustrations that did far more than document a deployment. They preserved identity. Humor. Brotherhood. Character.

The subtle culture of a small team operating in high-risk environments. In those drawings, I recognized the team as it truly was. Not the formal version presented in reports, but the human version that existed between missions—men who worked closely, trusted completely, and endured the pressure of uncertainty together.

Pete's illustrations captured that reality with a clarity that photographs often fail to achieve. They remain, in many respects, one of the most authentic visual records of that era within UDT-13. Later, Pete extended that work through contributions to Fire in the Hole, the in-house publication of the UDT-SEAL Museum community, including the Winter 2000 edition. That publication represented something important to all of us who served. It was not official history in the conventional sense. It was an internal memory—written by those who lived it, for those who understood it. Within that context, Pete's work became part of a larger effort to preserve not just events, but identity itself.

That effort culminated in his commissioned combat mural at the entrance of the *National Navy UDT/SEAL Museum* in Fort Pierce, Florida. The mural stands at a threshold, both literal and symbolic. It is the first thing one encounters upon entering the museum, and in many ways, it serves as a visual introduction to the history of Naval Special Warfare. It is not decoration. It is narrative rendered in color and form.

The mural traces the evolution of the Frogmen and SEALs across generations—early Underwater Demolition Teams, riverine operations in Vietnam, and the continued evolution of maritime special operations forces. It reflects adaptability as a defining characteristic of our community, not as theory but as lived necessity.

Adaptability was not an abstract concept for us in Vietnam. It was survival. We operated in environments where terrain, water, and visibility constantly shifted.

We learned to function in uncertainty, to adapt without hesitation, and to rely on small-unit cohesion under extreme pressure. Pete understood that this adaptability was not just tactical. It was cultural. It defined who we were.

One passage attributed to Captain William M. Shepherd, USN, Navy SEAL and astronaut, captures this better than any formal explanation:

"*We are at home in the water. Frogs are a very fitting symbol for us. They are above all adaptable… It is our strong sense of team and respect for our teammates that defines us.*"

That statement reflects a continuity of identity that extends across generations of service. It connects riverine warfare in Vietnam to modern Naval Special Warfare operations, and ultimately to environments far beyond Earth itself.

That continuity became even more visible in the story surrounding Pete's mural dedication. When the mural was unveiled at Fort Pierce, Captain Shepherd was not physically present. At that time, he was serving as commander of the International Space Station, orbiting Earth aboard Space Station Alpha. His presence in orbit while being honored on the ground created a symbolic alignment that was difficult to ignore. It represented a community that had extended itself beyond traditional boundaries—land, sea, and even space.

From orbit, Captain Shepherd transmitted a message back to Fort Pierce. In it, he reflected on the meaning of Pete's mural and on the identity of Naval Special Warfare. Capt. Shepherd spoke of adaptability, of operating in extreme environments, and of the defining importance of the team. He described his experience serving alongside fighter pilots, test pilots, astronauts, and cosmonauts, and noted that despite the diversity of backgrounds, the SEALs he had known remained distinguished by one constant: their ability to function as a cohesive unit under any condition. He closed his message with the phrase: "*Commander sends HOOYAH.*"

And added a postscript that became part of our shared history:

"*Pete has made this thing big enough we can just about see it from space.*" There is humor in that statement, but there is also truth embedded within it. Pete's work was not large in the physical sense alone. Its significance comes from what it represents. It preserves the lineage of men who operated in silence, often without recognition, and frequently under conditions where success itself required invisibility.

When I consider that mural today, I do not see paint or composition alone. I see continuity. I see Vietnam, where men like Billy Machen made the ultimate sacrifice under my command. I see the evolution of a community that learned to adapt across decades, technologies, and operational environments. I see the transition from riverine warfare to special global operations, and ultimately to the recognition that the principles governing success remain unchanged. The obligation to ensure that those who served are not reduced to footnotes in operational history.

Pete Carolan gave us a means to fulfill that obligation. Not through formal documentation or statistical recordkeeping, but through something more enduring. He preserved memory in visual form. He gave identity to experience. He translated the intangible into something that could be seen, studied, and remembered by future generations. In doing so, he became more than a participant in our history. He became one of its custodians. That is what endures in Naval Special Warfare. Not simply what was accomplished, but what is remembered. And in that sense, Pete's work stands at the entrance not only of a museum, but of memory itself.

<u>FULL MESSAGE BELOW:</u>

1000Z 12 NOV 00

FROM: SPACE STATION ALPHA

TO: UDT/SEAL MUSEUM, FT PIERCE FLA

SUBJ: MURAL

I WANT TO SHARE WITH ALL OF YOU SOME THOUGHTS ON THE DEDICATION OF PETE'S MURAL AT "MUSTER." I AM GREATLY HONORED.

IT MAKES ME THINK A LOT ABOUT THE TEAMMATES I SERVED WITH AND WHAT MADE THEM SPECIAL. PETE HAS CAPTURED A KEY ASPECT OF BEING A FROGMAN—WE ARE AT HOME IN THE WATER. FROGS ARE A VERY FITTING SYMBOL FOR US. THEY ARE ABOVE ALL, ADAPTABLE. AS SEALS WE TRANSITIONED TO ANOTHER ENVIRONMENT AND EXCELLED. I AM VERY PRIVILEGED TO BE MAKING ANOTHER ADAPTATION NOW, IN ANOTHER ENVIRONMENT—IN SPACE.

IT IS OUR STRONG SENSE OF TEAM AND RESPECT FOR OUR TEAMMATES THAT DEFINES US. IT IS FITTING THAT THE MUSEUM PROVIDES A MEANS TO DISPLAY THESE TRAITS FOR THOSE WHO WANT TO UNDERSTAND US AND TO KNOW THE HISTORY OF WHAT WE HAVE DONE.

WE ARE OFTEN BERATED BECAUSE OUR ROOTS ARE SIMPLE AND WE FAVOR WHAT WORKS OVER WHAT IS HIGH-TECH. SOME FIND IT SURPRISING THAT SEALS HAVE A PLACE AT THE FRONT OF THE SPACE PROGRAM. THERE HAVE EVEN BEEN SPEECHES IN THE RUSSIAN DUMA ON JUST THIS SUBJECT. I HAVE FLOWN WITH FIGHTER PILOTS, TEST PILOTS, TOP GUNS, COSMONAUTS AND MEN WHO WALKED THE MOON. I CAN TELL YOU AS FACT THAT SEALS ARE THE BEST OF THE BEST AND THIS IS BECAUSE WE KNOW HOW TO BE A TEAM.

I DON'T THINK THERE IS ANY REAL LIMIT TO WHAT SEALS CAN DO.

WE ARE 2 RUSSIANS AND AN AMERICAN, AND WE HAVE A BIG JOB AHEAD OF US—BUILDING THE REST OF SPACE STATION ALPHA WITH 16 COUNTRIES. NOT LONG AGO OUR SOYUZ PILOT WAS SITTING ON ALERT IN HIS MIG 21 ON THE BLACK SEA. AS WE FLY 200 MILES UP IN SPACE, I SHOW HIM THE PLACES AROUND THE WORLD WHERE I WORKED AND TRAINED AS A SEAL. WE HAVE A GREAT TEAM ONBOARD—ONE I WOULD HAVE BEEN PROUD TO SERVE WITH IN ANY SWIMMER LINE OR PATROL. I THINK ABOUT HOW FAR WE HAVE COME . . . NOT JUST AS A CREW,

BUT MAYBE AS A NATION.

I'M NOW PUTTING TO GOOD USE ALL THE THINGS I LEARNED FROM THE EXCEPTIONAL LPO'S, CHIEFS AND O'S THAT I SERVED WITH IN THE TEAMS. IN THAT SENSE, I WOULD HOPE THE TEAMMATES WHO ARE MUSTERED TOGETHER TODAY FEEL A PART OF THIS GREAT ADVENTURE WE ARE ON—AND THAT THEY, TOO, HAVE A HAND IN THIS SPACE HISTORY THAT WE ARE NOW WRITING.

COMMANDER SENDS.
HOOYAH

POSTSCRIPT—"PETE HAS MADE THIS THING BIG ENOUGH WE CAN JUST ABOUT MAKE IT OUT FROM SPACE."

CLASS 47' BUD/S GRADUATING CLASS

Front Row, Left to Right: Brian Kelly, Ted Mosure, Mike Sandlin, Hugh "Hughie" Freel, Tom "Rocky" Wagner, James Riter, Loren Decker, Russell Martin, Moses Simonet Marquez, George "Rusty" Skidmore, Mel Hardman.

Back Row, Left to Right: Edward Saldana, Morey Kucich, Pete "the Pirate" Carolan, Dave Carlton, Bill Bruce, Joseph Quincannon, Dale Bosley, Charles "Sandy" Prouty, Roger Clapp, Melvin "Skip" Crane, Robert MacFarlane, Udomporn Sompongse, Eric Letkeman, Robert "Robbie" Clay, Mike Kearns, George Hudak.

Front Row L to R:

1. **Brian Kelly:** He grew up in the greater Los Angeles area. When he arrived for training, he said he completed eight hundred push-ups daily to prepare. His arms and shoulders confirmed that. He served in UDT-13. When the instructors asked him to teach us a song, given his Irish heritage, he taught us "Molly Malone." He had three older sisters and said that the Beach Boys used to practice at their home, which I understood to be a good-sized area. Brian said that they went there to impress his good-looking sisters. Brian's sons play in a popular band in the San Juan Capistrano area ("The Kelly Boyz"). They have performed for the UDT/SEAL reunion at least once.

2. **Ted Mosure:** To make a little extra money in training, I bought hair clippers and offered haircuts to the guys on Sundays so they could pass inspection the next day. I learned as I went along. We all started with bald heads. Possibly the slowest swimmer in the class. Running and the Obstacle Course was better. Diving and demo classwork were easier. Deployed with Skip Crane's Kilo Platoon in July 1969 and returned to Vietnam in August 1970 as an

LDNN Asst. Advisor. Finished enlistment on the SEAL Basic Indoctrination staff (SBI).

3. **Mike Sandlin:** Retired years ago, after serving about 30 years. He was part of the training unit staff. Among his awards, he earned a Bronze Star with "V" device (valor) for action in Vietnam. He was inducted into the Ohio Military Hall of Fame years ago, as he was born in Hamilton, Ohio. He helped our group in honoring Class 347 in 2022 upon their graduation. He served with UDT-13.

4. **Hugh "Baby Hughie" Freel:** Also known as "Good Deal Freel" in circles, he was originally from Tawas City, Michigan. I have heard him described as a Force of Nature, or "Larger than Life" for his unending enthusiasm during training. Periodically, the boat Freel continued, crews would have "chariot races" with instructors riding atop the IBSs (inflatable boat small) as trainees ran along the beach. Freel would commonly be heard shouting something like, "We need more instructors!! This boat is too light!" He kept everyone laughing and spirits up as we went through the grind. Regretfully, he passed away from throat cancer in 2008. He served 26 years, achieving the rank of Senior

Chief before retiring in 1993. He served in UDT-13. While in Vietnam, he organized and ran the LDNN (Vietnamese SEALs) training camp at the Navy base in Cam Ranh Bay. After Vietnam, he worked for a civilian contractor, supplying the training area out on San Clemente Island with food and other supplies. He and his wife, Jenny, were high profile residents of Coronado, hosting countless social gatherings. Our family stayed at their home during one vacation back in the 1980s. Hughie, along with Rocky Wagner, were voted Co-Honor men for Class 47.

5. **Tom "Rocky" Wagner:** Voted Co-Honor man with Hugh Freel. Wagner's father, Lou Wagner, was on active duty as a UDT member in Coronado. He grew up in California. He was an exceptional runner, breaking the 4-minute mark in high school. He was assigned to SEAL Team One and deployed to Vietnam at least once. He now lives with his wife, Susan, in Northern California.

6. **James Riter:** He was one of several rollbacks from Class 46 after we completed Hell Week. I did not know him well. He was quiet and focused on the task at hand. He grew up in the Phoenix, AZ area. He was killed along with Frank Bomar in Vietnam.

7. **Loren Decker:** Loren was from Waco, Texas. He attended college for several years in Kearney, Nebraska. He went through training as an enlisted man. He served in Vietnam with Rocky Wagner in Charles Prouty's platoon, SEAL Team One. After his enlistment, he finished college and re-entered the Navy as an officer. He served more than 20 years in the Navy. He was highly intelligent and had a profound sense of humor. He composed several songs that we would sing during training. Again, some were rated "R" and others were to be sung where no one else could hear them.

8. **Russell Martin:** He was assigned to SEAL Team One. He was from Oregon. A good swimmer, running was difficult for him. He was quiet and a diligent worker. When he was on the SEAL teams, he let me borrow his car to drive to Lompoc, CA, for the weekend. My uncle noted the tires were marginal and bought a complete

set for the car. Russ saw that after I got back and told me I could borrow his car anytime!

9. **Moses Simonet Marquez:** Moses and I were in boot camp when we passed the UDT Screen Test on the same day. I have known Moses longer than anyone else on the SEAL teams. He was a gentleman and a strong trainee, putting in his utmost effort in everything he did. He was assigned to SEAL Team One and deployed with Skip Crane's Kilo Platoon, along with me. After being injured in Vietnam, he rotated back to the U.S. he served two tours of duty in Vietnam. He stayed in the Navy, retiring as a Senior Chief Hospital Corpsman. He is a great friend.

10. **George "Rusty" Skidmore:** Also known as "Squatty Body," Rusty was my swim buddy. He completed training as an E-4 while most of us were E-2s or E-3s. He managed his leadership role well. Running was difficult due to his muscular legs, but he put in maximum effort and performed well in every other aspect of training. He had a good sense of humor. Originally from Oregon, he was assigned to SEAL Team One.

11. **Mel Hardman:** Mel was our oldest trainee and served as a Radarman 1st Class (E-6). He was 29 years old. I cannot recall if he was assigned to UDT or SEALs upon graduation.

Back Row, L to R:

1. **Edward Saldana:** Originally from Kansas, he did everything well during training. One remarkable anecdote: He got married to his wife, Linda, the Saturday before Hell Week started the next day. After Hell Week was over, it was quite an adjustment. A couple of years later, I had the honor of being their daughter's godfather at her baptism in Coronado. Ed deployed with Charles Prouty's platoon along with Decker and Wagner. He did not stay in the Navy after his enlistment was over. He returned in 2022 with Crane, Carolan, Marquez, Sandlin, and Mosure to attend the Class 347 Graduation ceremony.

2. **Morey Kucich:** Physically, the best trainee in the class. Excellent at running, swimming, O-Course, and anything else. He also deployed with the Kilo Platoon in July 1969. He was a reliable operator. He was on the U.S. Navy

Military Olympics Team (CISM), composed of SEAL and UDT members, who competed internationally against teams from NATO countries. During that era, the U.S.A. team was dominant. Kucich voluntarily left the SEAL team prior to his enlistment expiring and returned to the regular Navy.

3. **Pete "The Pirate" Carolan:** An excellent swimmer whose unique talent was illustration. He kept the class amused with cartoons and caricatures. Originally from the Huntington Beach/Newport Beach area, he was an Ocean Lifeguard prior to enlisting in the Navy. He was assigned to UDT-13 as the Cartographer. While serving with UDT, he was part of the recovery team for the Apollo 13 spacecraft. He generously donated his artwork talent, which is prominent at the UDT/SEAL Museum at Fort Pierce, Florida, and many other venues.

4. **Dave Carlton:** Another Ocean Lifeguard from the Los Angeles area, Dave was an excellent swimmer. He had a laid-back attitude and a good sense of humor. He would attend reunions regularly. After his enlistment, Dave became connected to the Hollywood scene and occasionally acted. In pre-training, he

"hypnotized" me on a Sunday as many of us were getting our gear ready for the next day. I remember the guys lifting me by my head and feet only. I was raised off the ground and kept my board-stiff posture.

5. **Bill Bruce:** Bill was a strong trainee. He served with SEAL Team One. I have not seen him since we were in the Navy. One incident I remember was at the very beginning of Hell Week, when he was a little late for "muster." The rest of us had just "hit the bay" and were neck deep singing songs. First-phase trainees wore green helmet liners at the time. Bill was being yelled at by the instructors. At a full sprint from the barracks, he launched himself at the edge of the sea wall, diving over many of us in the water below. In mid-flight, he calmly removed his green helmet liner, tucked it under his arm, and entered the water with a smile on his face. We all cheered!

6. **Joseph Quincannon:** Mr. Quincannon had an East Coast accent, was intelligent, and had a good sense of humor. I remember seeing him at UDT in 1976 when I returned to Coronado on vacation. He told me that I looked like I could "still do the job, which I appreciated. He

eventually achieved flag rank due to his strong leadership skills.

7. **Dale Bosley:** Mr. Bosley was my boat crew captain. He was married while in training, which must have been stressful. In later years, he oversaw operations at the Supreme Court Building in Washington, D.C.

8. **Charles "Sandy" Prouty:** SEAL Team One. A Naval Academy graduate, Mr. Prouty served in Vietnam, leading a platoon that included Rocky Wagner, Loren Decker, and others. He was awarded a Silver Star for action in Vietnam during that tour. He had a lengthy career with the FBI. He mentored a young woman from Columbus, Ohio, who attended the Naval Academy after I informed him of her acceptance. He retired as a Navy Captain.

9. **Roger Clapp:** SEAL. He served two tours in Vietnam, I think. Asst. The Platoon Leader then led his own platoon. Thoughtfully worked hard. Retired as a Navy Captain. Served as the UDT/SEAL historian for a good while.

10. **Melvin "Skip" Crane:** SEAL Team One. Class 47 Senior Trainee, LTJG at the time. Skip

worked with PBRs (Patrol Boat, River) in Vietnam, prior to entering UDTRA. After Class 47 graduated and those assigned as SEALs completed Army Airborne School at Fort Benning, GA, all attended the six-week SEAL Basic Indoctrination (SBI). Not long after, Skip formed the Kilo platoon and organized pre-deployment training. Much of that was conducted near Niland, CA, where we often used live ammo. Several classmates from #47 joined Skip when we deployed in July 1969: Mosure, Marquez, Kucich, and Hudak. Experienced SEALs were included in that platoon as well. Kilo Platoon had an active tour. Skip supervised the LDNN (Vietnamese SEAL counterparts) program in Vietnam on his next tour. He retired after more than 20 years in the Navy and continues to lead Class 47 reunion efforts.

11. **Robert MacFarlane:** Mr. MacFarlane was an impressive trainee, doing well in every aspect of training. With a gymnastics background, he could really rock through PT, especially pull-ups. After training, he was assigned to UDT, I believe. He was another thoughtful, introspective leader in our class.

12. **Udomporn Sompongse:** Mr. Sompongse was one of three foreign trainees who started with the class and the only one to graduate. He served in the Thai Navy. After completing UDTRA, we believe he returned to Thailand to set up a similar Naval Special Warfare component. It is likely that his proteges were involved in the 2018 Thailand rescue, which saved the young scouts trapped in flooded caves. I remember Mr. Sompongse smiling continually, even when the going got tough.

13. **Eric Letkemann:** Eric was a unique trainee in that he started with Class 44 and met difficulties with the instructor staff after almost completing the UDTRA course. He was given the opportunity to start from Day One with Class 47 and took advantage of it. Eric completed two Hell Weeks and every other aspect of training. He had a good attitude, kept us laughing, and was a solid trainee. He was assigned to UDT-13. I remember him having a dog in 1971 that was a mix between a Basset Hound and a German shepherd. "Scotus Delicta" looked just like a German shepherd with noticeably short legs… very funny. Eric returned to the Denver CO area after his enlistment. I have not seen him since 1971.

14. **Robert "Robbie" Clay:** He was a quiet gentleman. I do not remember him being angry or upset with anything. He was a good friend of James Riter. He was assigned to UDT-13 after training.

15. **Mike Kearns:** Mike was a cartoonist in his own right. MAD Magazine ran a series called Spy vs. Spy, and Mike would draw versions of these characters in his illustrations. He and Pete Carolan shared some of that talent. If Mike were a character in the "Odd Couple," he would be Oscar. He was fun to be around. He earned the name "Sea Slug" after an incident while we trained on San Clemente Island. Our assignment was too free-dive fifty feet from the ocean surface and brought up evidence that we got to the bottom. Mike produced a sea slug, which startled him after he realized what it was! He was assigned to UDT-13. He is another one I have not seen since 1971.

16. **George Hudak:** George grew up in Phillips, Wisconsin. He was maybe five'4" and wrestled in high school, finishing second in his weight class. We called him "Hudak" to differentiate him from George Skidmore, whom we called "Skid" or Skidmore. Hudak was a particularly

good runner and managed any of the physical requirements of UDT Training. He had to work on his swimming early on but later became the swimming instructor at the Coronado municipal pool. Hudak served for many years in the Navy and was promoted to Warrant Officer Four. He owned a vintage sailboat on which he took my family cruising San Diego harbor.

Recollections about Class 47:

1. Instructor Staff included Dick Pearson, Master Chief Terry Moy, Chief Allen, Chief Rose, Chief Perry, Joe Messenger, Vince Olivera, Bud Jureck, Tommy Hatchett, Doc Heatherington, Doc Cline, and Mr. Otte.

2. Consensus was that we started with roughly eighty-five trainees, which was reduced to fewer than forty by Friday morning, week one. By the time Hell Week started, we had twenty-eight. As I recall, of the twenty-seven graduates, twenty-two were original. With training now totaling 54 or 56 weeks, I asked Class 347 graduates from 2022 how many were originals. No one started with the class.

3. We had three foreign personnel. The two besides Mr. Sompongse were Mr. Yee, a Lt. in the South Korean Navy. He gave up within 2-3 weeks. He did not understand why "we loved it" when the going got tough. Lee Van Manh was an E-5 from South Vietnam. He quit during Hell Week
(week 5).

4. When I arrived for training after taking boot camp leave, Class 46 had just started a few days earlier. That gave me time to get into shape and acclimate to the routine. There were several men who were assigned to Class 46 but needed work on one or more aspects of training: Loren Decker, George Skidmore, George Hudak, Mel Hardman, a career E-3 named McMains, Ben Kema, and Russell Martin. All except McMains and Ben Kema completed training. After Class 47 had Hell Week, four more from Class 46 joined our class: Quincannon, Clapp, Ke-arns, Riter, and Clay. All these men completed training as well.

5. We had two trainees who would have made it through but were affected by unfortunate circumstances. One was a guy from Texas named Owens who sustained a severe fracture

of his femur during Hell Week. He stayed in the Navy but never reentered training due to that injury. Another individual, Bowman, also from Texas, was found to have a level of color blindness that prevented him from distinguishing green from red. SEALS use directional signaling devices with those colors. Both Owens and Bowman were good trainees and good guys.

6. The San Diego Tribune ran an article on "Hell Week" while we were going through it in August 1968. I still have copies of some of those articles.

7. The Navy was producing the recruiting film titled "Men with Green Faces." Our class was filmed during the first week of training, doing PT. That segment appears early in the film.

WITNESS TO THE WORK

MIKE SANDLIN SOC, USN, (RET.)

My name is Mike Sandlin, and I offer this chapter to "*The Painted Frog*" based upon my personal service alongside Pete Carolan during the formative years of what would become modern Naval Special Warfare. The purpose of this written account is to document my firsthand observations of Pete Carolan's artistic abilities, his contributions to operational documentation, and his early design concepts related to insignia that would later influence the visual identity of the Navy SEAL community.

I first met Pete Carolan in June of 1968 when we both reported to the Naval Special Warfare Center to begin training in what was then known as UDTRA (Underwater Demolition Training), which later evolved into what is widely known today as BUD/S training. BUD/S Class 47 was forming, awaiting the start of actual training. Class 44 was just finishing their training and were getting ready to graduate. Their class was returning to the barracks late (after midnight), excited about graduation, and rather loud and jubilant. One of the class 44 members, Lance Farmer was his name, was making noise, and Pete yelled for him to *"Knock it off. We are trying to sleep, asshole!"* Now, that was exactly the wrong thing to say to Lance. His response was immediate: "WHO SAID THAT?" Silence followed, and then Lance said, "WHICH ONE OF YOU WANTABEES SAID THAT TO ME?" Silence in the barracks, you could have heard a pin drop.

I could see Pete was not going to own up to his blunder; he was the smart one. So, being quite sure of myself, I said: *"I didn't say it, but it sounds like a good idea."* Lance then invited me outside, and I groggily accepted. Big mistake! I did not even see it coming. Lance nearly removed my nose from my face. As I remember it, my fellow BUD/S class 47 candidate, Ted Mosure, drove me to the clinic to have my nose sewn back on. I never saw Lance after that; he went on to serve honorably with SEAL Team ONE. I did, however, get to know Pete Carolan quite well. We were classmates and were in the same boat crew for *Hell Week* about 8 weeks later.

That is how my association with Pete "the pirate" Carolan began. From the moment we began that training cycle together, Pete stood out among the men in our class. Like all of us who arrived there, he had physical strength, endurance, and determination. However, Pete also carried with him a unique gift that very few in our ranks had—an exceptional talent for drawing and visual storytelling.

In the early days of training, long before any of us knew exactly where our careers would take us, it became clear that Pete saw the world differently from most of us. While many of us focused solely on the physical and tactical demands of the program, Pete had the ability to see and interpret what he saw in ways that translated naturally onto paper. He carried a sketchbook with him often, and whenever there was a moment of quiet between training evolutions, Pete could be found drawing.

These drawings were not idle doodles. Even in those early days, they reflected a remarkable understanding of form, proportion, and symbolism. Pete had an instinctive ability to capture both the physical environment and the spirit of the men around him. Looking back, it seems clear that his artistic instincts were already beginning to merge with the culture and identity of Naval Special Warfare.

After completing our training, Pete and I were assigned to Underwater Demolition Team 13 during the Vietnam era. This assignment placed us in one of the Navy's most specialized operational units. The work of UDT swimmers needed both physical courage and technical precision. Our missions often involved hydrographic reconnaissance along hostile coastlines, where swimmers entered unfamiliar waters under dangerous conditions to gather intelligence about underwater terrain, obstacles, and beach gradients that could affect amphibious operations.

During these reconnaissance missions, swimmers collected measurements and observations that were critical to operational planning. However, raw data alone was not sufficient. That information needed to be interpreted and transformed into exact visual representations for commanders and planners to understand the underwater environment. This is where Pete Carolan's talents became invaluable. I personally saw Pete take the measurements our swimmers gathered and convert them into detailed operational charts. These charts illustrated underwater contours, sandbars, obstacles, depth gradients, and shoreline approaches. Pete's drawings allowed our teams to visualize the terrain beneath the water in ways that simple numbers and notes could never convey.

What made Pete's work extraordinary was not simply the accuracy of his drawings but the clarity with which he presented complex information. His charts were both technically precise and visually intuitive. A commander looking at one of Pete's drawings could at once understand the physical layout of a shoreline approach and the potential hazards involved in landing operations. Pete was not simply an artist working from a distance. He was also one of the strongest swimmers in our unit. I saw him personally take part in reconnaissance efforts, running the sounding line offshore to ensure exact measurements. He often volunteered for the most physically demanding tasks associated with gathering that data.

After those missions were completed, Pete would sit down with the team's notes and measurements. With patience and precision, he would begin translating that information into visual form. Watching his work was remarkable. He would begin with light lines mapping the contours of the seabed and gradually build the image layer by layer until the underwater terrain appeared almost three-dimensional on the page. Those drawings were more than artwork. They were mission tools. They allowed our team leaders to make decisions that could decide the success or failure of an operation, and, in many cases, the safety of the swimmers involved.

Pete's artistic ability did not go unnoticed within the command. His reputation as both a skilled Frogman and an accomplished artist began to spread throughout the unit. Whenever there was a need for illustrations, visual documentation, or design work, Pete was the man people turned to. One of the most memorable examples of Pete's artistic contributions occurred during our deployment when he collaborated with LTJG Pete Upton who was in Japan to produce the command, Cruise Book. In military units, a Cruise Book serves as both a historical record and a cultural snapshot of the team during a particular deployment. It documents the men, the operations, and the experiences that define a unit's identity.

Pete's artwork appeared throughout that book. Every section of the publication included illustrations drawn by Pete. These drawings captured the spirit of the team in ways that photographs alone could not. He portrayed the life of UDT swimmers with humor, authenticity, and an understanding that only someone who had lived that experience could possess. His illustrations ranged from lighthearted depictions of life within the unit to more serious representations of operational activities. In every case, his drawings reflected both his artistic skill and his deep connection to the men he served beside.

However, one memory stands out above all others in my recollection of Pete's artistic work. At one point during our service together, Pete showed me a small spiral notebook he carried. Inside that notebook were numerous sketches and design ideas he had been developing. Among those sketches were several concepts for insignia designs that he believed could represent the growing community of SEAL operators. At that time, the SEAL Teams were still relatively new. The identity of Naval Special Warfare was evolving, and there was no universally recognized insignia to symbolize SEAL operators' capabilities and heritage.

Pete had been asked to explore what such a symbol might look like during a contest being held to create the special warfare device. I clearly remember him turning the pages of that spiral notebook and seeing five unique design concepts he had drawn. Each incorporated element standing for strength, maritime tradition, and warrior spirit. Pete had carefully considered how different symbols might come together to create a single emblem that captured the team's identity. What impressed me most was the level of thought behind those sketches. Pete was not simply drawing decorative images. He was exploring the deeper symbolism that might represent the character and mission of Naval Special Warfare.

One of those designs stood out. While I cannot claim that the final insignia adopted by the Navy was an exact replica of Pete's sketch, I can say with certainty that one of the drawings I saw in that notebook strongly resembled the elements that would later become part of the Naval Special Warfare insignia.

Pete later told me that his design had eventually been submitted for him through official channels to the Department of the Navy for consideration. According to Pete, when the design was reviewed and refined, only minor modifications were made. The most notable adjustments involved the direction of the flintlock pistol and the orientation of the eagle's head. These changes were made to make the overall figure appear more aggressive and forward-facing, projecting a stronger visual presence when worn on the uniform.

Aside from those refinements, Pete indicated that the core elements of the design remained consistent with the concept he had originally sketched in his notebook. When I reflect on those early drawings, what strikes me most is that Pete was thinking about the identity of Naval Special Warfare long before it had fully solidified. His sketches were attempts to capture the spirit of the teams—the combination of maritime skill, airborne capability, and warrior ethos that defined SEAL operators. Pete was not only documenting the team's culture through his artwork; he was also helping shape how that culture might be symbolized. Years later, when the Naval Special Warfare insignia became widely recognized, I often thought back to those pages in Pete's spiral notebook. I remembered the care with which he had drawn each design and the excitement he expressed when discussing what the emblem might represent. From my perspective, Pete Carolan's contributions went beyond simple artistic expression. His drawings helped preserve the visual history of the teams during a period when Naval Special Warfare was still developing its identity.

He was both a participant in that history and a chronicler of it. In addition to his operational charts and insignia concepts, Pete's artwork captured the humor, camaraderie, and resilience of the men who served in UDT and early SEAL teams. His illustrations showed the human side of a community that was often defined only by its operational achievements. Those drawings remain an important record of the culture and spirit of Naval Special Warfare during that era.

Pete Carolan was a Frogman, a teammate, and a man who served with courage and dedication. But he was also something more—an artist whose talent allowed him to capture and communicate the essence of the teams in ways that words alone could not. From the charts he drew to support operational planning, to the Cruise Book illustrations that documented the life of the unit, to the insignia concepts he developed in his notebook, Pete's work reflects a unique intersection of art and military service.

I consider it an honor to have served alongside him and to have witnessed his creative process firsthand. It is my sincere belief that Pete Carolan's artistic contributions deserve recognition as part of the broader history of Naval Special Warfare. His drawings not only supported our missions in the field but also helped shape the visual identity of the SEAL community. This chapter and my official statement are offered based upon my personal observations and my service alongside Pete Carolan during that formative period.

THE FROGMAN AND THE FINAL CANVAS

TED MOSURE BUD/S CLASS 47
SEAL TEAM 1

The combat mural artwork, known worldwide, is always on my mind when I remember Pete Carolan. This mission to tell his legacy is to make sure that the world knows him not just as a Navy SEAL UDT 13 Veteran who served two tours in Vietnam, but as the artist who captured the spirit of the frogmen. During the Vietnam era, the term "frogman" was our badge of honor. It referred to the Navy's elite special forces, and our work was vital to the war effort. This project, this book, is about cementing his reputation while simultaneously raising critical funds for veterans' mental health programs through the SEAL Heritage Foundation, Pete's foundation, which he began and left to us all to carry forward. It is a final, vital act of brotherhood. I often think back to the first time our paths crossed. It was at Coronado, CA, during the relentless, soul-crushing early phases of BUD/S. Pete stood out at once to me.

He was an ocean lifeguard, a natural-born waterman, moving through the freezing bay with an ease that just floored me. I was the opposite, a native Ohioan with no competitive swimming or saltwater experience. Although I passed the UDT Screen Test comfortably, swimming in the ocean was new to me. I respected Pete's ability in navigating the ever-changing currents of the Pacific. Our bond was forged through BUD/S fire, right there in that cold, relentless environment, recognizing in each other the raw determination to survive the impossible.

Our shared journey began with BUD/S Class 47 class, one of the fourth classes of six that started the 18-week training in 1968. That shared grit is what defined us.

My path to becoming a Navy SEAL was not a childhood dream; it was a necessary decision against the backdrop of the Vietnam War draft in 1968. I looked at the options and chose the most demanding one: the Navy's Underwater Demolition Team (UDT) program. UDTs, which originated in World War II to clear beach obstacles, adapted quickly to the riverine warfare of the Mekong Delta. If I were to serve, I would do so with the most elite unit I could find. After graduation from BUD/S, I was assigned to SEAL Team 1.

I took the screen test during boot camp and was one

of only three candidates to pass, entering BUD/S at 19. I was not a natural athlete, and Pete had me beat by a mile in the water. But I had conviction. The program was designed to break you down, that was the point. We started with around eighty-five trainees, and by Friday morning of the very first week, fewer than forty remained. Success was not about brute strength; it was about sheer perseverance and finding a solution to the immediate, impossible task. We learned about the bell, which dreaded brass bell standing on the quarterdeck. It is more than just a piece of metal; the bell is the option to quit. To walk away, ring it three times, and be done with the cold, the pain, and the terror.

There was no bell to ring for our class. The procedure was to inform the instructor(s) during the evolution (task) that you did not want to continue. Quitters removed their helmet liners and reported to the instructor's office. The lessons, especially from men like the legendary BUD/S instructor, Terry "Mother" Moy, taught us that true strength was not just physical. Moy, with his five Vietnam tours, embodied the core philosophy: the courage was not in quitting, but in having the internal fortitude to solve the problem—to overcome the agony—so you did not have to stop. That kind of mental toughness, that relentless commitment to purpose, was what they instilled in us, and it was the necessary foundation for the elite Frogmen the Navy needed. I developed a profound respect for that process because it revealed the core of who we were.

The first training phase was a blur of constant motion and bone-aching cold. The relentless inspections, the unending push-ups, the miles of running, and the signature, punishing chilly water swimming in the bay. Most swims were in the ocean, and yes, it was still very cold. We were a global group, too, with foreign trainees like Mr. Yee, a determined Korean lieutenant. Mr. Yee went through everything we did, at least into week three. The LDNN program was conducted overseas and provided indigenous Vietnamese Navy personnel with an abbreviated version of our training in the U.S.—a small taste of the training role UDTs also performed, specifically with the Lien Đoan Nguoi Nhai (LDNN), or "Vietnamese Frogmen."

Through the exhaustion, humor was a lifeline—the dark, ridiculous comedy that kept our minds from fracturing. I will never forget a morning inspection by Terry "Mother" Moy. The pressure was intense. I decided to pull a stunt just to break the suffocating tension. I grabbed a gallon of mustard and used it to hide behind the candidate who was seen during inspection just the day before with mustard on his collar. As we all lined up, I hid a large jar of mustard behind the candidate, and when Moy came around the corner to see this, I expected immediate hell. Instead, it led to a moment of chaotic, shared laughter—even from the instructors. It was a moment of grace.

My punishment was swift but shared: I joined Russell Martin, consuming copious amounts of mustard, then spent the next hour in the freezing bay "washing" it off our uniforms. "*Hit the bay*" was the command as we ran to leap into that bay in full dress combat, fresh from inspection. Similar memories of shared discomfort, where comedy was the primary escape, are what Pete and I always bonded over. It perfectly encapsulated the spirit of the Teams, where the distinction between UDTs and SEALs was not always clear—we trained and operated together as "frogmen."

Pete's artistic talent was not limited to just cartoons and sketches. His ability to capture a moment, an emotion, or a personality was incredible. This artistry found a bizarre and unforgettable outlet when he served on the capsule recovery team for the Apollo 13 mission. Amid the national crisis and the pressure to bring the astronauts home, Pete, always leaving his mark, painted an image of "Freddy the Frog" on the Apollo recovery aircraft. "Freddy" was a swaggering, cocky cartoon frog with a determined attitude who wore a sailor cap, chewed on a lit cigar, and carried a lit stick of dynamite.

It was classic Pete: a flash of irreverent humor and quiet pride placed right on the stage of world history. The frog emblem was the insignia for the Helicopter Anti-Submarine Squadron 4 (HS-4), the unit responsible for the primary recovery of the Apollo 13 crew. The helicopter that retrieved astronauts Jim Lovell, Fred

Haise, and Jack Swigert were operating from the prime recovery ship, the USS *Iwo Jima.* The Famous Scene: In the 1995 film *Apollo 13,* directed by Ron Howard and starring Tom Hanks, the helicopter landing near the capsule and the subsequent retrieval of the astronauts are accurately depicted.

My training progressed steadily. I graduated from BUD/S in November 1968. While Pete went to UDT 13, I stayed on base with SEAL Team 1 for specialized training. The SEAL teams, created in 1962 from UDT frogmen, were focused on advanced land combat tactics for guerrilla and counterinsurgency operations. We were known for clandestine, lethal missions—the enemy often called us "the men with green faces" because of our camouflage. I went on to Fort Benning, Georgia, for airborne training, and later entered a six-week SEAL Basic Indoctrination course (SBI) in early 1969.

Upon completion of SBI training, I was assigned to the Kilo Platoon, a group of two officers and twelve enlisted personnel, led by Lt. Skip Crane. Kilo platoon, like other platoons, conducted extensive pre-deployment training over the next several months. During my four-year enlistment, I completed two six-month combat tours in Vietnam. With Skip Crane leading many of the SEAL team missions, Kilo platoon conducted over one hundred combat operations in six months during that first tour. Skip was a dedicated, focused leader, and I was committed to the platoon and our mission. I, along with other members of the Kilo platoon, became very competent and effective as we conducted our operations against Vietcong leaders.

After my first tour, I was assigned to SEAL advanced training and selected to serve as an advisor to Vietnamese SEALs as part of the Vietnamization program. Upon returning from Vietnam, I completed further training and was assigned as an instructor in the SBI program for the remaining time of my enlistment. The integration of the UDT and SEAL roles became complete years later, but our shared "frogman" legacy was sealed in Vietnam.

The Painted Frog is, without a doubt, the defining symbol of Pete Carolan's legacy. Today, his combat murals—huge, evocative pieces of art—are a testament to that legacy, featured in hallowed locations like the Pentagon and the White House, outside the Navy SEAL UDT Museum in Florida, and across the legendary military bases at Coronado, California. Pete's artwork, the tangible proof of his unique service, is everywhere. The SEAL brotherhood is a lifetime commitment. We maintain a steady friendship, seeing each other at reunions throughout the years. And Pete always showed up with his art. He would dedicate a piece and sign it to express his appreciation for the service.

His murals were not just paintings; they were a profound expression of our shared world. They reflected the camaraderie, teamwork, and resilience forged at BUD/S and in Vietnam. We know that some of his work is even related to intelligence efforts. His dedication to capturing the moment was absolute. Even recently, he created a powerful piece for Class 347 in 2022, the year he passed away. He was truly a combat muralist, using his talent to capture the experiences and the enduring spirit of the Teams. What we can all learn from Pete is that the strongest warriors are often the ones who find a way to express and process the chaos they face, transforming it into something enduring.

Now, our final mission for Pete is to ensure that his artistic legacy endures and helps his brothers. I pulled out a book he had recently signed for me, complete with personalized drawings. It is an invaluable piece of history that we plan to preserve, and it is invaluable to my family and me. Pete was a unique force—a tough frogman with an artistic soul, who constantly looked to connect and leave a meaningful mark. His commitment to the teams never faded, and even now, long after his passing, that dedication continues to serve a vital purpose.

We learned in training that failure was never an option, and the mission was always paramount. For many veterans, the wounds carried home are not visible, and the battle for mental peace is the hardest one they will ever fight. Pete understood that deeply, which is why he founded *The Patriot SEAL Heritage Foundation,* which highlights mental health services for our veterans and first responders.

This book is the final, most important action to

ensure that the Trident under which he served and the memory of Pete Carolan continue to serve those who follow in his wake. The funds generated by his art and this story will not be spent on stone monuments, but on the living, breathing needs of our community, supporting the veterans' mental health programs that save lives every single day. Pete's painted frog and his powerful murals remind us that the spirit of the frogman is not just in the fight—it is in the bond we share, the healing we look for, and the enduring commitment to leave no one behind.

THE UNSINKABLE SPIRIT OF "PETE THE PIRATE"

CHARLIE "SANDY" PROUTY
BUD/S CLASS 47 / SEAL TEAM 1

Executive Assistant Director of Law Enforcement, FBI (Ret.)

Every Frogman who knew him has a Pete Carolan story. They usually start with a punishing run, a near-drowning, or an instructor yelling until his voice breaks. My own story is no different; yet the lesson I learned from fellow BUD/S candidates like "Pete the Pirate" Carolan defined my subsequent three decades, culminating in my role as the Executive Assistant Director of Law Enforcement at the FBI (Ret.). My career in federal law enforcement, from a street agent to the Bureau's command staff, was not built in Washington, D.C.; it was forged in the sand of Coronado, under the watch of men who had seen the worst of war and guided by a peer whose spirit simply refused to submit. That lesson, shouted over the perpetual roar of the Pacific, was simple: "Forget how to quit." It was an ethic that transcended rank and branch, becoming the very bedrock of our lives.

When I reflect on the weight of the Executive Assistant Director title—the oversight, the politics, the impossible crises—it feels insignificant compared to the primal, necessary courage Pete embodied when facing a relentless surf at zero-dark-thirty. This chapter is dedicated to that indomitable spirit, the man who brought levity to Hell Week and who left us a legacy of compassion as enduring as any Trident.

We came to Basic Underwater Demolition/SEAL (BUD/S) training from two profoundly different worlds, underscoring the Navy's elite's wide-net approach. I was a Naval Academy graduate and an Officer. After fulfilling my required year aboard a ship, where the most demanding physical activity was often the climb up the gangplank, I arrived at BUD/S in a body that was, frankly, not in decent shape. My muscles had atrophied from long hours in the wardroom, and the program's physical toll was immediate and brutal. I was a fish out of water on the beach. Pete Carolan, by contrast, was the enlisted man, a natural force of nature.

He was a star athlete and swimmer, one of the fastest swimmers I had ever seen, his physique honed by his time as a supreme lifeguard at Huntington Beach and the disciplined intensity of Naval boxing. Pete was precisely where he was meant to be—in prime shape, radiating a raw, powerful, aquatic confidence. He carried no physical deficit; he was a machine built for this crucible. Yet, even Pete, the paragon of fitness, was humbled by the ceaseless pace. It was a terrifying moment for the class when this aquatic titan, this image of physical perfection, momentarily faltered. That such an intense, punishing regimen could bring even Carolan to his knees proved that the program was designed not just to test strength, but to break the will. It was the greatest psychological equalizer.

The only genuine, soul-saving relief in that intense, sleep-deprived psychological pressure cooker came from Pete. It was Carolan who gave the "best pep talks," words that had nothing to do with naval tactics or mission planning and everything to do with making a heavy situation light. He earned the moniker *"Pete the Pirate"* for his relentless, high-spirited defiance—a wise guy who found humor in the collective misery, often right under the noses of men who breathed fire.

Pete's talent for creating hilarious character depictions of the instructors was a lifeline for the class, a small, shared rebellion that kept our spirits from snapping. He captured the hyper-masculine attitudes and distinct physical characteristics of men like Pearson, the unflappable leader; "Mother" Moy, the unholy master of discipline; Vince Olivera, and Chicken Chest Joey, giving them all nicknames that stuck like mud.

These men were our mentors—the toughest instructors in the military, Vietnam veterans who were building us into elite fighters—and we owed them everything. But if Pete were tweaking the instructors with his brilliant, sarcastic humor, the entire team would have to do more running, lifting, swimming, push-ups, you name it. The extra attention was universal and total, yet we endured it. We gladly paid the price for the laugh because Pete was fundamentally unbreakable, and his ability to inject humor and light was the most powerful morale booster we had.

The true testament to our collective resolve, the final, painful articulation of the Frogman brotherhood, came at graduation. This was the ceremonial throwing of the class leader into the bay, the "Bendix," an act of defiance against the institution itself. Phase Leader Terry "Mother" Moy—a genuine legend with five combat tours in Vietnam and an unprecedented 5-term BUD/S instructor—was the target, and he warned us against it. Spurred on by Pete's mischievous defiance, the class's collective nerve was at its peak.

Despite the fear, I knew someone had to go first. I took the lead, running full steam at him, determined to be the first one to engage and commit the class to the action. Terry, true to his formidable reputation, met me head-on. The impact was like hitting a brick wall made of steel and leather; he struck me square in the jaw, nearly breaking it on contact. For a terrifying second, the world went black, but I stayed on my feet, anchored by the momentum of the class surging behind me. The job was done. The physical cost was immediate and searing, but the momentum was set, and with the class behind me, we ensured Terry "Mother" Moy went into the Bendix. At that moment, the pain, and the triumph—was the culmination of every shared misery: a painful final handshake with the instructors and a testament to the unbreakable unity of BUDS class 47.

After graduation, our paths diverged into the brutal reality of the war effort, each assignment reflecting the skill set forged in the Coronado surf. Pete Carolan—the brilliant, enlisted man—went into specialized reconnaissance. He served two foundational tours in Vietnam as a cartographer with UDT 13. This was not a desk job; it involved swimming miles of river and coastline at night, under the enemy's noses, mapping critical waterways and infrastructure with meticulous precision. It was the silent, unseen foundation upon which all kinetic operations depended.

My own role as a Naval Academy Officer led to immediate combat leadership. I received orders to lead a SEAL Team 1 platoon in Vietnam at once after earning my Trident, and I executed combat missions in the field. My experiences as a SEAL Team 1 leader were the direct continuation of the BUD/S oath, translating resilience into the command responsibility of young men's lives. A weight I carried every day and night in the Mekong Delta.

The intensity and unique commitment Pete showed in the rivers of Vietnam were called upon at once when he returned for a mission of national significance. It is still one of the most significant pieces of Frogman history: Pete was part of the NASA Apollo 13 capsule recovery team in 1970, an event of global significance that captivated a world in fear. The contrast was stark transitioning his UDT skills from navigating enemy waterways to retrieving astronauts from the vast, open Pacific. His irrepressible artistic signature, the infamous character *"Freddy the Frog,"* was brought along for the ride.

Pete painted Freddy onto the Sea King helicopter used for the recovery, an unofficial mascot of hope and grit that flew out to meet the returning crew. This helicopter, now permanently preserved by the UDT-SEAL Museum Association in Fort Pierce, Florida, stands as a testament to his talent. The work was so iconic that the actual helicopter was featured in the movie Apollo 13, starring Tom Hanks. The man who could make us laugh while running through the surf was the same man who added a touch of Frogman defiance and humor to one of the most serious national emergencies of the era, leaving a powerful mark on history and popular culture.

The training and the Frogman code stayed with me when I joined the FBI in 1973. I carried the "no quit" mandate into every assignment, from serving as an investigator in South Carolina, Virginia, and

Washington, D.C., to my time at Headquarters, focusing on applying the small-unit tactical intensity of the SEALs to complex criminal and national security crises. I went on to become a Plank Holder of the Hostage Rescue Team (HRT) in 1983, helping to set up the Standard Operating Procedures for the Bureau's elite tactical response unit—a direct application of the combat lessons Pete and I learned in Vietnam and Coronado. Later, I served as a supervisor of the HRT and the Violent Crimes section. Following September 11th, Director Mueller entrusted me with a crucial task: I was appointed the Executive Assistant Director of Law Enforcement in 2002.

This high-level position was established post-9/11 to confront terrorism and improve communication with state and local police. In all those roles—from forming HRT's structure to leading organizational transformation under the immense pressure of a post-9/11 world—I thought of Pete. Every time resilience was mandatory, every political and professional crucible, I heard the Pirate yelling: *"Is this the best you got?"* That spirit was the fuel that allowed me to endure the years and lead the Bureau's core law enforcement functions.

While I pursued law enforcement, Pete Carolan embarked on a final, two-fold mission that perfectly encapsulates the "NO QUIT" ethos, trading his weapon for a paintbrush. First, he became a world-renowned combat muralist. His influential, detailed art is not confined to typical galleries; it functions as visual military history and sacred space.

His work hangs at the highest levels of American defense and honor—permanently displayed at the Pentagon, throughout the U.S. Navy SEAL Museums and Military Bases, and even at the White House, commissioned during the era of Former President Bush.

His murals grace the places where veterans gather in Coronado, CA, serving as emotional anchors that visually recount the sacrifice and brotherhood of the Frogmen, ensuring the stories are passed down. Pete's art provides a profound voice to the silent service, capturing the gritty reality of the training and the quiet valor of combat in a way no photograph or official history ever could. He transformed his ability to make us laugh into an enduring visual language of honor and remembrance, creating monuments that will outlast us all.

When Pete Carolan passed away in August of 2022, the Frogman community lost one of its most vital and joyful voices. Yet, he left behind his most incredible charitable legacy, the *Patriot SEAL Heritage Foundation.*

This foundation is the ultimate realization of the Pirates' humor and humanity—a deep understanding that the most formidable warriors who never quit on the outside often need the most compassionate support for their hidden battles to fight the silent war with PTSD, TBI, and related conditions long after the uniform is folded and put away.

THE EAGLE WATCHES: BROTHERHOOD THAT ENDURES

TOM "ROCKY" WAGNER BUD/S CLASS 47

SEAL TEAM ONE

I met Pete Carolan the same way a lot of us met the men who would shape our lives forever—Coronado, CA, where the Pacific does its best to break you and the instructors finish the job. Navy SEALs BUD/S training, inside of Class 47' is where our story begins.

Back then, those numbers did not mean much to anyone outside of training, but to us, they became everything. That was the proving ground. The place where titles did not matter, where your past did not carry you, and where every man had to answer the same question: *Are you going to stay, or are you going to quit?*

I did not get there by accident. I came into the Teams during the Vietnam draft, following in my father's footsteps. My dad, Lou Wagner, was part of UDT 11—one of the early frogmen. There were a hundred SEALs total on the West Coast back then. Tight community. Tight brotherhood. When I joined, we became the first father and son to serve together at the same time in the Teams.

But let me be clear about something—legacy does not buy you anything in BUD/S. You must earn every inch. When I found out I was going to be drafted, I made a call to a man named John Steph, someone already in the Teams. My parents were stationed in Okinawa at the time, and I needed help navigating the process. I knew where I wanted to go. I was not interested in anything halfway. If I was going in, I was going all the way.

I had one thing working for me—speed. Back in high school, I was the first recorded to run the mile in under four minutes. That kind of conditioning does not just happen—it builds something inside you. A tolerance for pain. A relationship with exhaustion. The understanding that your body will give out long before your mind must. That helped me in BUD/S. But it did not make it easy, nothing makes it easy. That is where I met Pete. I knew he was a champion swimmer, and I could feel the respect his trainee candidates had for him in the water.

Now, Pete was not the loudest guy out there. He did not need to be. Some men try to prove themselves every second of the day—Pete did not. He had a quiet way about him, but you could feel it. There was something steady there. Something grounded. And when things got hard—and they always got hard, Pete lit up in a unique way. You could see how much he belonged to that environment. Some men fight the grind and others become it. Pete became it.

We were all shaped in those days—physically, mentally, and in ways you do not even understand until years later. Training was not about the water and our first phase instructors such as Vince Olivera, Master Chief Terry Moy and our proctor Dick Pearson drilled that into us. These were men who were tested by fire and the respect we had for each of them, it shaped us for battle.

Some of the most important work happened on land—learning pursuit, evasion, ambush. Real-world scenarios that would translate directly into Vietnam. I remember one evolution that stuck with me. Our class was training to join the conflict in Vietnam with border patrols with Mexico. Charlie "Sandy" Prouty—another brother who became a permanent part of my life and who I served two tours within Vietnam was on patrol during a night of training. We practiced real life scenarios that could be the dividing line of life and death while on patrol in Vietnam. During that exercise, we encountered what was called a "coyote." Back then, that term referred to individuals smuggling people across the Tijuana border for money.

This was not theoretical; this training was real. The "coyote" ran and I chased him. I did not think about it. I did not hesitate. He took off, and I locked in. It turned into something more than a drill—more than training. It became instinct. He cut through terrain, trying to shake me, and I just kept going. I ran four miles into Mexico chasing that man. Four miles without looking back, I knew my brothers were on surveillance ready to assist.

Did I catch him? No. But that is not the part anyone remembers. What they remember is that I did not stop. That is the mentality BUD/S builds. You do not quit because it is inconvenient. You do not quit because you are tired. You do not quit because you are outmatched. You go, and then you go some more. At that moment I stayed with the guys. It became one of those stories that gets told and retold—not because it was heroic, but because it represented something we all understood. Never quit, never ring the bell.

When I think back on our training, running was not easy. My fellow class 47 BUD/S trainee Mike Sandlin, and I used to run together. Long runs. Hard runs. And there was something we all understood without saying it: no one finishes last. If a guy started falling back, I would drop back with him matching his pace, because we all finished together. That is how we survived. That is how we operated in combat. That is brotherhood. And when that goes away—or when it changes—it can hit a man hard. I have seen it in others as well. That look some guys get—the "death look." It is not physical. It is internal. It is what happens when a man is trying to figure out where he fits once the uniform comes off.

Pete understood that better than most. That is the thing about the Teams—what you build there does not go away. It does not fade. It becomes part of you. To this day, when I talk about Charlie "Sandy" Prouty, I say, *"I'm still his shadow."* And I mean that. That is not nostalgia. That is brotherhood. That is what BUD/S does when you go through it the right way.

Pete carried that same connection—but in his own way. Inside the Teams, Pete was alive. There was a light in him when he was around the guys, when he had a mission, when he had purpose right in front of him. But when that chapter started to close—like it does for all of us—there was something for which he was searching. Not purpose but connection to a brotherhood.

He was trying to recreate something that only exists in one place—the brotherhood we had in those Teams. And Pete found his way to do it. Through his combat military art that depicted the super patriots of our special warfare forces. One of his most famous pieces is the mural at the entrance of the *National Navy SEAL UDT Museum* in Fort Pierce, FL. It is clear to all who visit the museum the enduring spirit of patriotism and pride for our United States special force teams. A time capsule to never forget.

Now, I remember seeing him years later, during his Reserve days, over near the Oceanside complex, across from the old base where SEALs still train. While a lot of guys moved on—different careers, different lives—Pete stayed close. Not in uniform but rather in spirit. He would be out there painting murals and walls of the base. And I am telling you—these were not just paintings. These were tributes. These were living pieces of history. Every brushstroke meant something. I would stand there and watch him work sometimes. I would tell him, *"Pete, this is something else."* He would smile at me, salute, and return to his work.

His work reminded me of the great muralists—like Wyland over in Hawaii—but Pete had his own thing. His own voice. I told people all the time, *"That's the work of an ace painter."* But here is the part that mattered more than the talent—Pete knew why he was painting.

He used to say, *"I'm here holding down the fort while our men are fighting."* Think about that for a moment. He did not see himself as separate from the mission. He saw himself as part of it—just in a different role. Those combat military murals were not decoration. They were remembrance. They were his connection to his country and brotherhood that stands the test of time. They were a way to say: *"You're still with us, we honor you."*

Pete had humility about him, but there were two sides to him. The quiet observer—and the man deeply rooted in his brothers. He did not need recognition, but he deserved it. One of the things he became known for by all of us who saw his creative genius firsthand was his influence on the eagle in the SEAL Trident. That eagle—it means something. To me, it always felt like more than a symbol. It felt like a message. Like Pete was saying, *"I'm watching over my flock."* That is who he was at his core. Always looking out and always connected to his service and commitment to his country.

And he had a way of embedding himself into his work. Pete had this habit in his sketches and murals. Like a "Where's Waldo" game—but instead of Waldo, it was him or his famous frogs. We would look at these pieces and say, *"Find Pete."* And sure enough, somewhere in there—hidden in the scene—was his likeness. A figure. A face. A presence. Much like his painting of "Freddy the Frog" on Helicopter #66 at the USS Midway Museum in San Diego, CA. He left his mark for his SEAL team brothers who made history safely bringing home our Apollo 13 astronauts in 1970.

That was his way of staying in it and of never really leaving. Because the truth is, for a lot of us, leaving the Teams is not a clean break. Some guy's struggle with it and find unhealthy ways to cope, his art kept him engaged in every healthy way possible.

That is the real thing, and it is not talked about enough. But Pete found a way through it. Art became his anchor and his foundation, *The Patriot SEAL Heritage Foundation,* works to create solutions for our veterans and first responders. The work they do to pair up vets with K9 training, and the three-day equine retreats are evidence that Pete's work goes on. His art therapy that he poured into to stay healthy and move his life forward is his legacy and his everlasting connection back to everything we had all built together.

And in doing that, he did not just help himself, he helped all of us. Because every mural, every piece, every hidden "Find Pete" moment—it reminded us that we were still part of something. That we still mattered and that the brotherhood did not end.

Pete was a super patriot. Even after his time in active service, even when he was in the Reserves, he carried that same level of commitment. He believed in the mission. He believed in the men he fought for and with and he believed that every little bit helped the cause.

That is something I carry with me even today. Something I think we all should. Because at the end of the day, beneath all the training, all the stories, all the legends, there is something simple we need to remember. There is a human being there. Behind every uniform. Behind every mission. Behind every story, Pete never lost sight of that. And that is why his work still resonates with the way it does. Because his art was not about war. It was about people. About brotherhood. About connection. About never quitting—not just in the field, but in life.

I still see him, you know. Not in the way you might think. But inside of art itself and the shared stories right here in *The Painted Frog*. In the way we talk about those days and the memories we share. And sometimes, when I look at one of his pieces, I catch myself doing it again, scanning the image. Looking closely. Smiling when I find him or his hidden frogs inside of all his artwork. Because he is still there. Still part of it. Still holding the line.

And if you take the time, if you really look...You will find Pete too and you may just find a piece of yourself. Support the foundation so that our United States veterans and Navy SEALs have a place to be seen, connected, and heal. A brotherhood that was forged in the fire, where the watchful eye of the eagle protects and serves, because it is there that his brotherhood endures.

THE DAY THAT THE TEAMS GOT THEIR SYMBOL

BILL POZZI, BUD/S CLASS 46 UDT 13 VIETNAM VETERAN

We had been back from the Vietnam War for a little while when this moment took place, and even now, looking back, I can still feel the atmosphere of that time war was raging and it was a stressful time to be in the Navy. It was not just the end of a war; it was the beginning of a complicated return into a country that did not quite know what to do with the men who had come back from it. Coronado carried its own version of that transition. The base was still active, still disciplined, still structured, but underneath it there was a kind of quiet uncertainty, the civilian population had great disdain for the military. If you had short hair, you were immediately called out. Coronado felt like a safe place, if you went into San Diego your windshield would be broken when you came out of a movie or a grocery store.

It was a Thursday when I saw Pete sitting in one of the classrooms drawing cartoons, completely absorbed in his work in a way that made the rest of the world feel temporarily irrelevant. There was always something steady about Pete when he was drawing or cartooning, like his focus created a boundary around him that kept everything else from intruding. I asked him what he was doing that weekend because, as I knew from habit, he usually made the drive up to Orange County to see his family, a rhythm that grounded him in a life outside the military structure we were all living inside.

He didn't look up right away, just kept drawing as he told me he was staying in Coronado because he had entered a drawing contest, something official, something coming from an entity that had issued a challenge to design a new badge for UDT and SEAL Team. At that time, before anything like that existed, the only insignia we had were simple and utilitarian, a shoulder semicircle with our unit designation and, if I recall correctly, a basic scuba badge that marked function more than identity. It told people what we were assigned to, but having nothing to do with the unit we were assigned to but not the specialized group of elite warriors.

I had known Pete before any of us were in the Navy. We were not close in a dramatic or defining way, but we were connected through the overlapping geography of youth. He swam and played water polo at Orange Coast College, I was at Compton College, and somewhere in those years our paths crossed, and we became acquaintances in that loose coastal way young men do when they recognize common traits in each other. We were both swimmers, water polo guys, and drank beer and chased women. We both lived in Huntington Beach at different points, moved through the same surf culture, knew some of the same people, and shared an unspoken understanding of that Southern California world before everything shifted into military structure and wartime experience.

Pete was serious about this project in a way that immediately stood out. It was not casual sketching or passing time. It had weight to it, even if the outcome was still uncertain. He showed me early samples he had been working on, including Freddy the Frog in different poses, and a quite simple SEAL concept that featured an actual seal—minimal and unrefined at that stage, but consistent in its effort to find form. Watching him work, you could see he was not just drawing images, he was trying to produce a symbol that would give us pride in the unit we belonged to.

That Friday after seeing him, I drove up to Los Angeles, and I remember thinking about his project the entire time, the idea of it staying with me longer than I expected. There was something about the possibility of it that felt larger than the assignment itself, as if what he was doing was trying to give shape to something that had never been formally seen, only lived. The Vietnam-era environment we were returning into made everything feel slightly out of sync. There was pride in service, but also a cultural distance growing between those who had served and the public that often did not understand or want to understand. We were sent to war by President Johnson and people did not care, they vehemently opposed the Vietnam war. They wanted the war to stop. We had learned skills that were activated and the frogman excelled in the war. The public hated us for following the orders of a President who trained us to protect and serve.

That tension existed everywhere, even in small interactions, even in silence. It shaped how we moved through the world outside the base. I saw Pete again after work on Monday, and by then something had clearly changed in his process. He had moved beyond early sketches and was now showing me twenty distinctive designs, each one more developed than the last. The evolution was visible immediately. The initial ideas had been simple, almost exploratory, but the newer versions carried intention. He had brought in a red Budweiser label, peeled carefully from a bottle, using it as a prototype background. On that familiar eagle backdrop, he began layering symbols that represented several aspects of what we did and how we operated. Some designs included fins, others an M16, an anchor, a Trident, a small boat resembling a Swift boat, and even a parachute. Each combination explored a different expression of capability and identity. It was not random; it was analytical creativity; a visual language being evaluated in real time. He also had sheets of insignia from the Marines, Army, and Navy spread out as reference points, not to copy but to understand how other branches had translated identity into emblems.

There was a seriousness in how he approached it that felt different from anything else happening around us at the time. Because the SEAL Teams were still so small within the Navy structure, I never thought something like this would ever be taken seriously. There were 3 UDT teams and 1 SEAL team on each coast. We all knew each other well. In my mind, we were too new, too specialized, too easily overlooked within the larger machinery of the Navy. The idea that an enlisted E-4 could design something that would become official felt almost impossible, not because of talent, but because of bureaucracy and tradition.

Pete submitted his work that week, and like most things that pass-through systems like that, we moved on and forgot about it. There was no expectation attached to it, no belief that it would resurface. It became just another small moment in the long continuity of returning to routine after war, where even meaningful things often disappear into administrative silence.

But about six months later, something happened

that none of us were prepared for. A prototype emerged that the Navy was proposing to adopt, and when I saw it, I felt a kind of disbelief that took a moment to settle. It was almost identical to what Pete had submitted. The reaction among us was not immediate celebration, it was quiet astonishment, the realization that something we had dismissed as unlikely had taken form. I remember thinking that the design itself was unusually bold in its clarity. It did not try to impress through complexity. It communicated directly, instinctively, what we were and what we did. It felt like one of those rare symbols that does not require explanation, like commercial logos that become part of cultural memory—Shell, Ford, Chevrolet, McDonald's—simple forms that carry meaning instantly because they are so distilled. Pete's design achieved something similar, a reduction of complex identity into something visually immediate and enduring.

When the finished insignia was finally issued to us months later during quarters in the afternoon, we were each handed the UDT device and told to wear it on our dress blues for inspection the next morning. There was a quiet shift in atmosphere when that happened, not dramatic, but noticeable, like something internal had been acknowledged officially. The next morning, we arrived for inspection and immediately realized no one was entirely sure where the device was supposed to be placed. Some wore it on the right side, others on the left, some above the pocket, others below. It revealed, humorously, how new the symbol still was to us. The XO Lieutenant Robert W. Peterson and Lieutenant Commander James B. (Jim) Wilson, our CO, eventually stepped in and clarified the correct placement, and we adjusted accordingly, aligning ourselves both literally and symbolically.

In those days, enlisted devices were silver, officers wore gold, a small distinction that reflected the structure we lived within, but the meaning of the emblem itself went beyond rank. Everyone was glad to have it. It created a sense of identity that had not existed before, something that belonged specifically to us rather than being assigned generically by the larger institution. It added pride in a time when pride was complicated. The public, at that time, did not hold service members in high regard. There was open hostility in many places, and those in uniform were often reduced to labels that stripped away individuality and experience. The phrase "baby killer" was something many of us encountered, directly or indirectly, and it created a kind of emotional distance between who we were and how we were perceived. Against that backdrop, having a symbol that stood for us internally mattered more than it might have in another time. It was not just a badge. It was recognition, identity, and unity. It said something about who we were becoming, even when the world outside misunderstood us.

Looking back now, I understand that what Pete created was never just an assignment or a design exercise. It was a translation of lived experience into form, an attempt to give structure to something that had previously existed only in action and memory. At the time, none of us fully understood the significance of it. We were too close to it, still moving through the experience of war, still adjusting to a world that felt slightly unsteady. But over time, the emblem became something more than we realized it would be. It became enduring. And in that endurance, it carried forward not just a design, but a moment in time when identity was still being created, by instinct, and by the quiet creativity of a man sitting in a classroom in Coronado drawing cartoons and concepts that would eventually define us in ways none of us could have predicted.

When Pete would sit to draw and cartoon, he would look up at times while drawing the device that tested his skill and manhood. He did not realize the significance of it all—creating a device from a beer label with an eagle, solidifying the feeling of the Teams. He believed he could, so he did.

I miss his lunacy, his assumptions with no basis in reality. He followed his instincts and would get into trouble after his service years and during his time in the reserves. Pete seemed to struggle with a reality that was not organized for him to follow orders. I excel in diplomacy and bureaucracy so I would coach him on how to handle himself outside of the Teams. Sometimes he followed, most times not. I admired him and Frank Sinatra; they did it their way. And some men, like Pete, were never meant to be corrected—only remembered for their contribution.

THE BROTHERHOOD BENEATH THE SURFACE

GIL "ESPI" ESPINOZA UDTRA CLASS 42

"The hardest thing was not Vietnam.
The hardest thing was not BUD/S-UDTRA.
The hardest thing was deciding who I was going to be when the pain came."
— Gil "Espi" Espinoza

When I sat down with Gil Espinoza, he asked me to call him "Espi." I expected to hear stories about Vietnam, SEAL training, and combat. I expected tales of hardship and brotherhood. What I discovered instead was a man who had spent decades searching for the meaning behind suffering, service, and survival. Espi's story intersects with Pete Carolan's in a way that helps explain not only who Pete was, but what made the frogman community so unique.

The two men came from diverse backgrounds, but they shared something forged in the same crucible—the unforgiving environment of BUD/S training and the even greater challenge of finding purpose after the war.

Espi graduated from BUD/S Class 42 during one of the most turbulent periods in American history. Vietnam was raging. Casualties were mounting. The nation was divided. Young men were being drafted, and for Espi, the decision to enter military service came after a deeply personal experience while he was wrestling on a full scholarship in Sterling, CO.

Espi met a friend whose brother had been killed in Vietnam, and the loss impacted him deeply. *"It affected me,"* Espi recalled. *"I couldn't stop thinking about him."* That tragedy became a turning point, and he enlisted. Rather than waiting for life to decide his future, he chose his own path and enlisted in the Navy to become a Frogman. His wrestling coach had been a Frogman; he set his mind to task and left college to fight for his country. He had no idea then how much the training would shape the rest of his life.

Many people describe BUD/S as a process designed to break men down and rebuild them. Espi rejects that idea completely. *"They didn't break me down to build me up,"* he said. *"The training revealed who I was."* That distinction is important. In Espi's mind, BUD/S was not about creating warriors; it was about revealing them. In BUD/S training and FIRE, the mission is designed to be perfect!

This perfection is implanted into our mindset, and

that is the level we operate from. When the mission does not go as planned, we can adjust, adapt, and improvise, which is a good thing; it is a time when our lines get blurred, and we function outside the lines, like the spray of a paint can. I see in Pete's ART that the lines are not linear, sharp, or defined; it is a lens into freedom. In our world, it is tough to allow ourselves to see ourselves outside of perfection, and we struggle. Pete allowed himself, in his art, not to be perfect; in life, he struggled (PTSD) as we all do. Pete expresses this struggle through shaded tones, blurred lines, and not-so-perfect scenarios, allowing us to accept our own imperfection.

When the chilly water in the San Diego Bay stripped away comfort and when exhaustion stripped away pride and excuses, what remained was the authentic man beneath the surface. The first-phase BUD/S instructors were not manufacturing character; they were uncovering it. The same could be said of Pete Carolan. Pete's paintings often revealed truths that words could not express. His frogs, seals, and military combat dream-like combat scenes were not merely artistic choices. They reflected a deeper reality that only those who had endured the training and combat could fully understand.

Espi immediately recognized that quality in Pete's work. *"There was something about his paintings,"* he said. *"Dreamlike, but direct. They connected emotionally."* More than once, Espi mentioned Pete's recurring imagery of frogs and seals together. To outsiders, they might appear whimsical, but to Frogmen, they represented brotherhood, a shared identity, and a bond forged through camaraderie. Simply put, a language that is understood without explanation.

During BUD/S, Espi trained under legendary instructors including Chief Dick Allen and Boatswain Mate First Class Vincent Olivera. The names of these two legendary BUD/S first and second phase instructors still carry weight among those who came through the program during that era, as the standards were relentless and the consequences for weakness were immediate. Yet Espi remembers something far more significant than the punishment; he remembers the men. *"The pain wasn't what stayed with me," he said. "The brotherhood stayed with me." THERE IS POWER IN THE PAIN TO TRANSFORM A TADPOLE TO A FROG.* That brotherhood became even more important when Class 42 graduated.

Unlike many classes before them, Espi and sixteen of his classmates were drafted directly into SEAL cadre training instead of serving UDT tours first. Espi talks openly about this experience on ***Jocko Podcast*** episode # 361. Jocko Willink is a retired U.S. Navy SEAL TEAM 3, Task Unit Bruiser, who focuses heavily on leadership, discipline, and military history. Espi tells how Vietnam needed operators, and so the pipeline accelerated under the direction of Captain Franklin Anderson.

The war was waiting, but before Espi deployed, fate

intervened. He broke his ankle during platoon workup and spent two months in the hospital recovering. In this recovery time at Balboa Hospital in San Diego, Marines from Khe Sanh came pouring in, and Espi saw the raw casualties and consequences of war stripped away his vision of war. For many young men, such a setback might have been devastating. For Espi, it became another lesson; he was going to war. He learned that perseverance often looks different from what is expected and that sometimes perseverance means charging forward and sometimes it means waiting. Eventually, he joined another platoon and deployed to Vietnam.

Combat brought realities that no training evolution could fully prepare a man to face. Espi shared with me memories of firefights, ambushes, and teammates who carried one another through impossible circumstances. Each memory carried lessons that extended far beyond military tactics. Repeatedly, Espi returned to a single theme in our sit-down conversation: we all make choices versus relying on circumstances, coincidences, or luck. He believes the choices made under pressure, when fear is present and when hope is absent, define a man's character. When you have already surrendered your will to something greater than yourself, fear loses its power, you stop worrying about survival, and focus on purpose, a freedom most people will never understand.

Years later, psychologists would call many of these experiences post-traumatic stress. Espi learned language through life experience through reevaluation counseling with the Boulder Fire Department after wrestling for the University of Colorado on a full-ride scholarship.

Like many Vietnam veterans, he had returned home carrying burdens he did not fully understand. He learned how unprocessed trauma can lodge itself deep within the human psyche and how memories can become trapped. How pain can remain alive long after danger has passed. Yet Espi bristles when people ask how he *"coped"* with PTSD. *"I don't like the word coping or surviving,"* he told me. To him, coping implies managing symptoms without addressing causes. It suggests endurance without transformation, and surviving is a word near victimhood; one cannot survive training to become a SEAL, one must prevail. Instead, Espi speaks about expression and releasing what has been buried and allowing the spirit to speak honestly.

His perspective mirrors something visible throughout Pete Carolan's artwork. Pete did not merely cope through painting. He expressed his emotions and translated them into color. He transformed memories into images, and he turned pain into creation. Espi believes that distinction matters. *"Pete wasn't coping,"* he said. *"He was creating."* The difference may seem subtle, but to Espi it represents two entirely different approaches to healing. He says, *"One manages wounds, and the other transforms them."*

The conversation eventually shifted toward faith. It

became clear to me that Espi's understanding of God had evolved dramatically over the course of his life. Growing up Catholic, he relied heavily on religious traditions and symbols. During Vietnam, he carried items in his pockets that he believed offered spiritual protection. Like many young warriors, he sought certainty in uncertain times. But after the war, a woman he knew challenged many of his assumptions about faith. He was introduced to the holy spirit, and what he encountered during his learning made him question and seek answers. The questions she asked forced him to examine what he genuinely believed. *"Who do you pray to?"* and for the first time, he separated religious doctrine from personal relationships.

Eventually, he embraced a deeper connection with Christ that transformed the way he viewed suffering. That transformation became central to his own healing. One verse he shared with me stayed with him throughout his life: *"Lean not on your own understanding but in all your ways acknowledge him and he will direct your path."* Espi explained that much of his pain came from trying to make sense of events that defied explanation. Such as:

Q: Why do some men live while others die?

A: *It is life! Death is common to all men, when and how is the big question. Death has no mercy. Time and chance* *guide the way to our end, or escape. I am in the right place and time to die, or I am not. Mistakes have brutal consequences, too; again, wrong place at the right time for Death to come. It is not the thing you see that kills you, it is the thing sniffing for your name.*

Q: Why does suffering exist at all?

A: *There is power in suffering to be transformed, to endure, persevere, to focus on something outside of yourself, trust in teammates and God, and know you will prevail in God's design for your purpose.*

These answers evolved, and he learned to trust, surrender, and to accept that understanding and faith are not always the same thing.

As he reflected on Vietnam and the times that surrounded that war, Espi spoke about moments of hopelessness and moments when quitting seemed easier than continuing. Moments when darkness felt overwhelming. Yet each time, he discovered something unexpected. Strength that arrived through connection, his teammates, through loved ones, and through God. He warns of isolation for those with PTSD. He told me, *"SEALs thrive because they belong to something larger than themselves." The mission matters. The team matters. The man beside you matter."* Those lessons remained with Espi long after the war ended.

In Vietnam, we would prepare to roll out, we would gear up, check our weapons, water, packs, and "paint up." "Green Face" was our color; each teammate would highlight their face in black. It is time to make War. I always saw this time as putting on my death face, the face the dying would see, and my death face, if death found me. It was my time of preparation, to operate at my highest level, time to call the "outside warrior," one of the Painted Frogs.

Years later, Espi became involved with veterans' organizations, including the Rocky Mountain SEAL Association. He remained committed to supporting fellow warriors navigating the challenges of life after service. His philosophy continues to evolve to this day. He says, *"Pain is unavoidable, misery is not. There is a difference between pain and passion. Pain consumes. Passion transforms. One drains life, and the other gives life meaning, and Pete understood that."* Espi believes Pete discovered peace through his art. Not because painting erased his experiences, but because painting gave him purpose.

The conversation turned toward Pete's contributions to the SEAL community. Espi recalled Pete as approachable, kind, and genuine. *"He made people feel seen,"* he said. It was not the first time in my interviewing of SEALs for *"The Painted Frog"* that someone had described Pete that way. Again, people mention his ability to make others feel important, to make them feel heard and valued. Espi believed that quality explained much of Pete's lasting influence. He says, *"People remembered how Pete made them feel. Not just what he accomplished. Not just what he painted. Not just what he built. They remembered the man, the friendship, the connection, and the authenticity that Pete brought to every mission and interaction he had with others."*

When discussing Pete's famous military combat artwork, Espi acknowledged that he was not personally present during every stage of the Trident design process. However, he confirmed Pete's involvement as part of the team whose artistic contributions helped shape the final insignia. He mentioned the feelings that came up for him while standing in front of the entrance of the National Navy SEAL | UDT Museum in Fort Pierce, FL, where Pete Carolan's combat mural welcomes visitors to the museum. *"Pete was not defined by a single achievement. He was defined by a lifetime of creating. A lifetime of serving. A lifetime of building connections. I do not view that mural as a symptom of trauma; I view it as evidence of peace. Pete looked like he was at peace with himself."*

That observation struck me while we spoke. Many people assume veterans create art because they are broken. Espi rejects that notion. He told me, *"Pete created art because he was whole enough to express what others could not, his art became a bridge between worlds. Between military and civilian. Between memory and understanding. Between pain and purpose."*

Toward the end of our conversation, I asked Espi what he thought veterans needed most. His answer was surprisingly simple. He stated, *"They need to discover their why. Not societies why. Not someone else's why. Their own. They must find in themselves the answers to questions like: Why did they live? Why are they still here? Why does their story matter?"*

He encouraged veterans to write those answers down to preserve them and to share them. Ultimately, leaving something meaningful behind for the next generations of warriors. The idea resonated deeply because it mirrors the purpose behind *"The Painted Frog."* This book is not merely about Pete Carolan; it is about preserving lessons, stories, and wisdom earned through sacrifice.

Espi's story begins in this chapter of reflection and legacy in his own life. He spoke openly about his desire to author a book for his grandchildren. After his house burned down and he lost everything in the last five years, he shared with me his reflection on loss. He felt this was a euphemism for burning it down and building it all up again. He is called to write for the next generation. Not a war book or a combat memoir. A book about service, faith, sacrifice, and discovering your purpose in life. In many ways, that desire echoes Pete's legacy. Both men understood that survival alone is not enough; the challenge comes afterward. The challenge is deciding what to do with the years you have been given.

As our interview concluded, Espi reflected on a phrase

often spoken to veterans. *"Thank you for your service."* His response surprised me. Instead of saying *"You're welcome,"* he often replies: *"You're worth it."* Those three words reveal everything about this man. The years, sacrifices, wounds, losses, and perseverance, none of them were meaningless. It was given in service to something greater. To teammates, family, future generations, our nation, and to his calling. And that is why Gil "Espi" Espinoza's reflections fit so naturally into Pete Carolan's story.

Both men understood that true healing is not found by escaping pain, but rather it is found by transforming it. Through brotherhood, purpose, service, and faith. And, through the courage to remain authentic when the world demands otherwise. That, Espi believes, is the lesson hidden beneath the surface of every Frogman's journey. Not survival, endurance, or even victory. But discover your authentic self when everything else has been stripped away and then choose, day after day, to live from that truth with faith.

THE PATRIOT SEAL

JAMES LAVORE , SEAL TEAM ONE
BUD/S CLASS 49

Here on Duck Key in the Florida Keys the storms continue to move across the water carrying the smell of salt and the ocean. Fishing boats rocking against the docks as the rain taps softly against their hulls before disappearing into the ocean. It is peaceful here. My thoughts often drift toward my Navy SEAL teammates. Pete "The Pirate" Carolan is always among them. To understand Pete, you should know this about him. Pete did not look at the world the way most people do. Where others saw obstacles, Pete saw something else. Solutions to problems mostly. He had the ability to focus on where he was going rather than what stood in his way. That quality followed him throughout his life.

During a West Coast SEAL team reunion, Pete invited my wife Kathy and I to Newport Beach to see his yacht, *The Patriot SEAL*. The way Pete talked about the vessel, you would have thought he was about to give us a tour of the USS MIDWAY aircraft carrier and museum. When we arrived at the marina, we could not find Pete anywhere. I called his cell phone, but there was no answer, so we waited. Then, from across the harbor I spotted a small inflatable dinghy making its way toward us. As it got closer, I could see it was Pete rowing it. The dinghy had the front chamber deflated and folded inside the bow. Water splashed as Pete worked the oars. He waved us aboard with a nod that suggested everything was going ahead according to his plan. Nothing rattled this former Huntington Beach lifeguard and former Navy frogman. Kathy looked at me and I motioned for her to get in the half-inflated dinghy. Pete rowed us across the harbor to "*The Patriot SEAL.*"

As we approached Pete's yacht, I could see how proud he was of that boat. Pete was going to bring her back to life from her grandiose past. Once we were all aboard, he took us below deck to show us the engine room. Pete spread his arms wide like a man unveiling a masterpiece. "*We're running on one engine this week Jim,*" he announced. I can still see him saying that even now. The engine room was a disaster in progress, but that was not how Pete saw it. Pete saw a yacht and it was his yacht! Only one engine was operating, but that was enough for Pete to get underway. "*One engine, as long as we have one engine, we're good to go!*" That statement captured something special about *Pete the Pirate.*

He did not spend much time thinking about what was not working. It was not that Pete did not want both engines to fire, it was that both engines were not needed to move his yacht around. After our tour of *The Patriot SEAL*, Pete took Kathy and I to Blackie's bar in Newport Beach for a beer. Anybody who knew Pete knew how much he loved that place. Years later, many of Pete's paintings would hang on the walls there, becoming part of the establishment's history.

We were sitting together at a high-top table when Pete's presence was requested by some young ladies at the back of Blackies bar. Pete knew everybody and the women loved him. While he was gone, a firefighter friend of his walked over to our table and began telling Kathy and I a story about Pete. There had been a fellow who regularly came into Blackies and made a habit of insulting military veterans, "a hippie", Pete called him.

Week after week, this hippie would show up with something negative to say about the men who had served our country. Pete returned to our table. The firefighter laughed and said, "*you tell them about what happened next Pete.*" Pete smiled and said, "*I threw him out the front window.*" He sat down and continued drinking his beer as though he had just described taking out the trash. The funny thing was that nobody seemed surprised.

That story captured the other side of Pete 's personality. People sometimes mistake his personality for rebellion, but I never saw it that way. Pete was not interested in fighting authority; Pete was interested in fighting injustice. There is a difference. If Pete believed someone was abusing authority, treating people unfairly, or disrespecting those who had earned respect, Pete got involved. He could not simply walk away. That quality had been with him for as long as I had known him. Our friendship went back to Coronado and the early days of Naval Special Warfare.

Pete graduated with BUD/S Class 47. I started in the same class, but a knee injury in the first phase rolled me back to Class 48 and then again to Class 49. I graduated with Class 49 alongside some exceptional teammates, including Medal of Honor recipient, LT. Michael E. Thornton. Back then the teams were part of our identity. Nobody talked about resilience because resilience was simply expected. You got hurt, healed up, and returned to work. Pete understood the mindset completely. When we had time off and enough energy left to enjoy it, Pete and I hitchhiked from Coronado to Newport Beach to visit his family. Those trips became part of a friendship that would last for decades. Pete stood apart from others. I could see that he respected competence and integrity, and he respected people who had earned their place. What he never respected was authority that relied solely on time in grade rather than competence, integrity, and character.

That trait surfaced again during an incident involving access to the Coronado Amphibious Base. Pete had a valid pass, but a disagreement arose with a gate guard. What should have been routine interaction became much larger. Eventually, Naval Special Warfare senior leadership became involved before the matter was resolved. Pete got his pass back, but the story was never really about the pass; it was about principle. Pete believed respect should travel in both directions. He gave it to me or anyone when it was earned and expected the same in return. This was one of the defining characteristics in Pete's life. Vietnam, for us, reinforced that attitude.

I deployed to Vietnam with DELTA platoon, SEAL Team ONE. On one of my sixty-five combat ops there, we were coordinating with a PBR unit and using the sound of the engine to help conceal our movement, we inserted and moved along a trail toward our objective. I was the point man, and I knew something was not right as we approached the village. My instinct had told me we were not where we were supposed to be. We had been inserted on the wrong side of the village. The pressure from our squad leader was to move faster, but I knew my instincts were telling me to do something different. Soon, I saw a villager, who was probably a Viet Cong Scout, coming down the trail with a lit candle in his hands. Moments later I would see a muzzle flash in my face. The night would erupt in gunfire. Muzzle flashes exploded in the darkness and rounds cracked through the air. Tracer fire streaked across the landscape. We were inserted in the wrong place, and this was only my second op in Vietnam. Pete and I stayed in contact long after the Vietnam conflict was over. We often spoke several times a week. The friendship never disappeared because friendships, as teammates in Naval Special Warfare, are different. Time becomes irrelevant; you simply pick up where you left off.

One evening, I was visiting a Newport Beach Police retired helicopter pilot friend in Montana. He began recalling an incident he had involving a Navy veteran. As details emerged, I found myself thinking about Pete. The description sounded too familiar, so I picked up the phone and called Pete to question him about it. When he answered, he told me it was in fact him and that he was angry about what was happening around him. Angry about situations he believed were unfair and angry about people who were not being treated fairly. When I put my friend Mike, the helicopter pilot, on the phone with Pete, we found out it really was Pete, and we all had a great laugh talking about what happened that night on a roof top in Newport Beach, California.

Pete carried many things with them throughout his life. Like many men from our generation of frogmen, he spent years around demolitions, explosive blasts, combat operations, and hard physical impact. Long before traumatic brain injury became part of the national conversation, an operator simply continued with the mission, and Pete was no exception. Pete as a

Boy Scout had fallen into a deep ravine and was found unconscious three hours later. In the 1950's, there was no technology to determine his type of brain injury, and so his injury went untreated.

Pete's paintings were not simply paintings to me; they were war stories. They reflected humor, patriotism, frustration, loyalty, history, and imagination all at once. A time capsule of memories for us all. The famous painted frog "*Freddy the Frog*" on Navy helicopter number 66 at the USS Midway Museum, during his NASA Apollo 13 capsule recovery became part of his mythology. Anyone who knew Pete understood that every piece of his art contained a little bit of the man himself. There is a hidden frog or K- bar knife in every painting he created. Pete "The Pirate" Carolan just saw the world differently; to me that was his gift.

As I sit here today in Duck Key, listening to the rain, I find myself returning to that image of Pete aboard *The Patriot SEAL* boat. I can still see him standing in that engine room, and I can still hear his voice. "*We've got one engine Jim, we're good to go!*" The older I get the more wisdom I find in those words. Life rarely unfolds exactly the way we planned. Boats leak, engines fail, and obstacles appear where we least expect them. Pete understood something many people never learned; keep moving forward anyway, and that was how he approached life. That was how he approached friendships. That was how he approached service to his country.

Pete Carolan spent his life standing up for what he believed was right. He defended veterans. He challenged injustice. He questioned authority when authority deserved questioning. He remained fiercely loyal to his friends and unapologetically devoted himself to the principles he believed in. Pete "The Pirate" Carolan was a super patriot!

Pete Carolan – San Diego Bay
Private Collection Photo – James Lavore

Apollo 13 Capsule Recovery – 1970 – @ NASA PHOTO

THE MEN IN THE WATER

There are moments in history that the world remembers in perfect clarity—where they were, what they felt, what was said. Apollo 13 is one of those moments. A mission that began as routine and quickly became a question of survival. An explosion in space. Oxygen is lost. Systems failing. Three astronauts who were drifting farther from Earth with each passing second, their fate uncertain.

On the ground, the world held its breath. Televisions flickered with the steady voice of Walter Cronkite, delivering updates that blurred the line between journalism and prayer. Families gathered. Strangers watched together. Nations paused. But far from the cameras and commentary, beyond the reach of the broadcast signal, another group of men waited. They were not in Houston. They were not in the control room. And they were not being briefed in real time.

They were in the open Pacific. Aboard the USS *Iwo Jima,* members of the United States Navy's Underwater Demolition Team 13—frogmen—prepared for a recovery mission they did not fully understand. While the rest of the world followed the crisis moment by moment, these men knew only fragments: there had been an accident in space, the lunar landing had been aborted, and the astronauts were coming home. That was all. No detailed updates. No unfolding narrative. No certainty about what condition the capsule—or the men inside it—would be in when it hit the water. And still, they prepared. Because that was their role.

Long before the world would come to know the Navy SEALs, there were the frogmen—quiet professionals trained to enter the most unforgiving environments on Earth. Their work began in the shadow of World War II, when Scouts and Raiders first cleared the beaches of Normandy under heavy fire, suffering staggering losses so others could follow. That lineage continued through the Underwater Demolition Teams, men conditioned to move through surf, darkness, and chaos with precision and calm. By the time of Apollo 13, they had become something rare: the bridge between space and sea.

When the capsule returned, it would not land on a runway. It would fall from the sky into an unpredictable ocean, where wind, current, and waves could turn even the most successful mission into a disaster. It would be these men—already in the water, already in motion—who would secure it, stabilize it, and make first contact with the astronauts inside. They would be the first hands extended back to Earth.

Among them were men like Pete "the Pirate" Carolan, Al "Smiley" Starr, and Lieutenant Ernest "Field Marshal" Jahncke. Different personalities. Different paths. But bound by the same training, the same code, and the same unspoken understanding: there is no room for error in the water. They did not operate as individuals. They worked as a unit. Trust was not discussed, it was assumed. Forged long before Apollo, in the brutal proving grounds of BUD/S training, where each man was responsible not only for himself, but for the life of the swimmer beside him. There, in chilly water and relentless conditions, a bond was formed that would carry into every mission that followed.

From the air, those in the command helicopter watched as the capsule cut through the sky on a razor's edge. Inside, astronaut Jim Lovell navigated with extraordinary precision, guiding the damaged spacecraft to a landing point so exact it would later be described as hitting a bullseye. From that moment forward, the mission belonged to the men in the water.

What followed would never be fully captured by cameras or commentary. The frogmen moved quickly—deploying, securing, stabilizing—executing a choreography learned through failure, adjustment, and repetition. There were no second chances. Only action. And yet, within that action, there were moments that would endure in unexpected ways. A glimpse of gold foil peeling from the capsule's surface as it met the ocean. Pieces of space itself, slipping loose, gathered instinctively and tucked into wetsuits. A quiet acknowledgment of something extraordinary, even as the mission continued. And then, on the deck of the ship, after survival had been secured and the world would soon erupt in celebration, there was an image waiting.

Painted on the side of the recovery helicopter #66. A frog airbrushed the night before recovery by Pete "the pirate" Carolan. To the outside world, it might have seemed like nothing more than a piece of military artwork. But to the men who had been there—to those who had trained, deployed, and carried out the mission—it meant something else entirely. It meant we were here. We did it. It was a mark. A signature. A quiet declaration left not in words, but in image. For Pete Carolan, that image would become something more. Not just a moment, but a legacy. A symbol that bridged who he was in the water with who he would become beyond it.

My sit-down interviews with Apollo 12 & 13 Capsule recovery UDT divers Al Starr & Lt. Ernst Jahncke are not just about the Apollo 13 recovery mission. It is about the men who stood in the space between crisis and return. The ones who were not watching history unfold—but were prepared to catch it when it fell from the sky.

These are their voices.

PETER J. CAROLAN

BUDS CLASS 47 | UDT 13 | APOLLO 13 CAPSULE RECOVERY 1970

MEDIA INQUIRIES: MEDIA@PATRIOTSEALHERITAGE.COM

JOIN US IN MAKING A DIFFERENCE

WWW.PATRIOTSEALHERITAGE.ORG

APOLLO 13

FREDDY THE FROG, CAPSULE RECOVERY 1970 AL "SMILEY" STARR

History remembers Apollo 13 as a story of crisis and triumph in space, highlighting the near-miraculous survival of three astronauts under circumstances that could have ended in disaster. However, what the world often overlooks rarely makes it into the headlines—is the critical work that took place on the surface of the Pacific Ocean. This is where men like "the Pirate", Pete Carolan and I operated as part of the NASA Apollo 13 Capsule Recovery Team. Our job was simple in description but heavy in responsibility: to secure the capsule, recover the astronauts, and execute every task flawlessly. There was no room for improvisation; yet, at times, it was precisely in those moments that ingenuity proved to be most essential.

The capsule splashed down in calm seas, which, on the surface, may seem like a blessing. However, calm waters can be deceptive. They reveal every detail, leaving no room to hide mistakes. Recovery operations require disciplined habits, experienced coordination, and the courage to act decisively in uncertain conditions. The astronauts' lives were in our hands, every stroke, every movement, and every decision mattered. Failure was not an option, and the margin for error was slim.

By the time I arrived at BUD/S training, Pete Carolan had already graduated from BUD/S. I joined afterward, and by then, he had earned a powerful reputation among instructors and trainees alike for being steady, dependable, and unflappable. Pete was not the loudest or flashiest person in the room, but that did not matter. The individuals who relied on his waterman's skill and presence rather than words already recognized his value. When I reached BUD/S, he had already proven himself as someone I could trust in any situation. That steadiness became our foundation when we were assigned to the Apollo 13 Capsule Recovery mission.

We flew to Hawaii and boarded the USS *Iwo Jima,* a Wasp-class amphibious assault ship of the United States Navy. We were awaiting our mission as the tension surrounding the NASA touchdown felt tenuous. The space mission was fraught with elements of danger, and the world was watching. The night before the capsule recovery, Pete approached me with an idea while we were in the sleeping bay of the ship, which, on the surface, seemed reckless. He had cut out three cardboard templates and carried them onto the ship like contraband. With a grin, he said, *"Let's go paint the bird that will lift us to the recovery of the astronauts."*

I stopped him immediately. Painting a military helicopter—one assigned to a NASA recovery mission—was not only against regulations; it could get us arrested. I reminded him of the consequences. I reminded him that even a momentary lapse could end both of our careers. The walk to the hangar bay felt interminable as I tried every argument I could muster. *"This is a bad idea,"* I said. *"We could end up in the brig. You are taking a huge risk. Think about this, Pete."*

Pete listened patiently, calm as ever. *"Everything will be fine,"* he said to me. *"You just hold the cardboard steady. I want to create a 3D effect with the airbrush cans I brought."* That was Pete—methodical, confident, unwavering. He was not reckless; he was intentional. Inside the hangar bay, we set to work. The metal walls hummed quietly under the ship's systems, and our footsteps were swallowed by steel. Pete laid out the templates, I carefully positioned them while keeping watch, and he began airbrushing layer by layer.

The image he created—Freddy the Frog—was more than a cartoon or decoration. It was precise, lively, and bold, full of presence, confidence, and subtle detail, and has since become an iconic military symbol for the Navy SEALS. As I held the cardboard in place for him, I understood what Pete had done instinctively: he had left a mark on history, on the bird that would carry astronauts from uncertainty to safety. By the time we finished, *Freddy the Frog* was alive on the fuselage.

We left the hangar bay quietly. No alarms, no confrontation. What mattered most had been done, and we went to our quarters to rest, unaware that the next morning would test more than our skill would test patience, nerve, and the unexpected flexibility of naval hierarchy.

At dawn, the discovery happened. The Commanding Officer saw the helicopter and went ballistic. Pete and I were summoned at once. He screamed about the destruction of government property, about insubordination, about how we were both being sent to the brig for pulling what he called a *"stunt."* He screamed that we would ride the ship all the way back to San Diego in just our swimsuits, under arrest. We stood at attention, taking it all in. That is what you do in situations like that. You accept accountability and wait for the consequences.

Then the Chief of Naval Operations arrived. He looked at the helicopter and smiled. He did not ask for an explanation. He asked if he could have a frog painted on his own bird. The air shifted instantly. The CO burned with disbelief, watching his authorities evaporate as the CO sanctioned what, moments earlier, would have been a career-ending offense. Both were commanders for the helo squadron. What Pete had created in the middle of the night was suddenly an approved symbol, and the frog became a quiet icon of ingenuity, morale, and pride.

The recovery operation that followed for Apollo 13 that day was intense, precise, and unforgiving. Pete and I boarded the helicopter, the very aircraft with *Freddy the Frog* painted on its side and lifted off toward the capsule. When the order came, we jumped into the Pacific, hit the water cleanly, moved fast, and secured the flotation collar on the capsule. The capsule was then stabilized.

The astronauts were alive. The helicopter lifted them from the ocean and returned them to the USS *Iwo Jima,* completing the recovery of NASA's Apollo 13 capsule. Every move was rehearsed for a month before the recovery, every action measured, yet nothing could ease the tension of knowing that lives hung in the balance in a mission that had held the world's attention. The SEAL frog went with us, visible to all, as a quiet reminder of the human element in operations that history often overlooks. Later, Hollywood would memorialize the recovery, with *Freddy the Frog* appearing in the film *Apollo 13,* starring Tom Hanks as astronaut Jim Lovell. Millions saw the image Pete painted without knowing the story, without understanding that the frog represented more than whimsy—it stood for presence, commitment, and the quiet audacity of men who operate in the shadows of history.

The helicopter now sits on permanent display at the USS *Midway Museum,* a tangible reminder of that mission and a visual testament to the men who executed it flawlessly. Visitors see the aircraft. They read the placard. Few realize that *Freddy the Frog* was never about rebellion or humor—it was about marking a moment of presence. About taking responsibility for a mission and leaving a symbol behind that would endure.

After Apollo 13 Capsule Recovery, Pete and I shipped off to Vietnam together to serve our country. That sequence matters. Recovery first. Vietnam next. Pete shipped out with UDT 13 as the cartographer, and the mission's intensity escalated from precise water jumps to patrols, demolition operations, and the ever-present stress of combat environments we all faced in Vietnam. The lessons from NASA and BUD/S—discipline, coordination, teamwork, and attention to detail—carried over directly into Vietnam. Pete's steadiness under pressure never wavered, and he continued to document what most people never see.

Pete Carolan's artwork from Vietnam captures moments of downtime, humor, fatigue, and quiet brotherhood. It does not glorify combat; it chronicles it. I am proud to own original pieces of his work—not because they are rare, but because they are authentic. When I look at them, I see memories accurately preserved. I see discipline paired with creativity, humor paired with resolve, and the human element behind every mission.

Pete never sought credit or recognition for his art or his service. However, history demands accuracy, and legacy requires that names be preserved. That is why *The Painted Frog* exists. This book records firsthand accounts, documenting the men who were there, the symbols they left behind, and the missions they executed without hesitation or complaint.

Pete Carolan is remembered not only as an operator but also as someone who recognized that symbols last longer than missions. The helicopter carrying *Freddy the Frog* is preserved at the USS *Midway Museum* in San Diego, CA, with a placard that reads:

> *Freddy the Frog – Apollo 13 Recovery Helicopter (1970). Painted by U.S. Navy sailor and combat artist Pete Carolan the night before the Apollo 13 capsule recovery, this iconic image became an unofficial symbol of resilience, morale, and ingenuity. The helicopter, now permanently displayed at the USS Midway Museum, was used to lift astronauts safely from the Pacific Ocean following one of NASA's most critical missions. Carolan and fellow recovery swimmer Al Starr both jumped from this aircraft to secure the capsule and crew. The frog later appeared in the feature film Apollo 13, cementing its place in American history.*

Pete's actions were quiet yet daring, and they had significant consequences. They marked the beginning of a legacy that would take him from the calm waters of the Pacific to the battlefields of Vietnam. His journey serves as a reminder that even within operations governed by strict rules, there is still room for creativity, courage, and individuality.

When I reflect on the Apollo 13 capsule recovery, I remember the calm waters, the jump, the collar, the lift, and that walk to the hangar bay the night before with my friend, who became my family and my brother. Pete was right, everything turned out fine, and a Navy SEAL frog made history.

The military combat murals and art are the legacy that Pete Carolan left behind for all of us: a blend of discipline and creativity, action and art, audacity, and responsibility. That frog did not just mark a helicopter; it represented a man, a mission, and a standard for those of us who followed.

Foundation Credits:
This chapter is dedicated to honoring the legacy of U.S. Navy recovery swimmers, Underwater Demolition Team (UDT) operators, and combat artists. Their contributions transcend formal records and are deeply embedded in operational history. We especially acknowledge the Patriot SEAL Heritage Foundation for its commitment to preserving untold stories, visual artifacts, and the lived experiences of America's elite naval warriors. Their efforts ensure that service, courage, and brotherhood are remembered in their entirety.

USS *MIDWAY MUSEUM* SAN DIEGO, CA. HELO #66
ARTIST *PETE "THE PIRATE"* CAROLAN
"FREDDY THE FROG"
UDT-13 | 1970 | APOLLO 13 CAPSULE RECOVERY TEAM

TRUST AMONG FROGMEN

LIEUTENANT (RET.) ERNEST "FIELD MARSHAL" JAHNCKE

Apollo 12 & Apollo 13 NASA
Capsule Recovery | UDT-13

Interview with Dr. (h.c.) Lisa Ray: 3/18/2026

Q: Hello, Mr. Jahncke. Thank you for taking the time to sit down and be interviewed by me today for the book, *"The Painted Frog."* I would like to open the conversation today and begin discussing both of your Apollo capsule recovery missions. The capsule recoveries were between Vietnam tours for you. As the Lieutenant of the NASA capsule recovery teams, how did you view the responsibility of leading men into a mission like the Apollo 12 and 13 capsule recovery missions?

A: *"I was on the mission of Apollo 12 capsule recovery. I* *then trained with those Apollo 12 astronauts Conrad, Gordon, and Bean. We trained in the water for the Apollo 12 mission, Corpus Christi, TX, in the Gulf of Mexico. I remember going out one morning on the CVS, the U.S.S HORNET, going through the whole recovery drill. The Gulf of Mexico was like a lake. Off Hawaii, it is blowing 30 to 40 knots in 10-12 feet seas.*

In our first attempts, one of my guys jumped in and could not get the sea anchor onto the capsule. We had not figured out how to go downwind of the capsule, and that first attempt was a disaster. Once we were back on the ship, humiliated, I was called to the officers' wardroom. There must have been what looked like a dozen Admirals, Captains, and other senior officers and the like, sitting and staring at me for answers. I was a young LTJG explaining how we could not get it done. I said, "We were not prepared for what we encountered. The Apollo 11 frogman left us no systems, procedures, or guidelines. We were starting from ground zero. Sea anchor and up, painful, timing jump on a crest of a wave, not a trough, which is preferable."

Q: Where both the 12 and 13 capsule recovery missions flawless?

A: *"Well… Apollo 12, I was supposed to open the hatch for the astronauts. The device I had, which looked like a key or a small wrench, would not open the hatch. This is one of my main jobs. As Walter Cronkite was speaking in* *real time on National News, Richard Gordon was looking at me through the window hatch inside the capsule. I was struggling to open the hatch. Luckily, the astronauts were able to open the hatch from inside.*

Apollo 13 recovery: safe and sound splashdown; no decontamination, as they never made a spacewalk. So, I decided I would do a 1.5-flip out of the side of the old number #66. I was last in the water, and no one saw that backflip except the commander of the helicopter squadron."

Q: Apollo 13 was not a routine capsule recovery—what was different about that mission from your perspective before splashdown?

A: *"The context was completely different. We knew less than almost anyone who was not aboard the USS Iwo Jima. The world was listening to Walter Cronkite, getting updates as they happened. But out where we were, in the South Pacific, we were isolated.*

All we knew was this: There had been an accident in space. They had not landed on the moon. And they were coming back. That was it. No additional details. No inside briefings. No special intelligence was passed down to the officers. Nothing beyond that. So, from our standpoint, we were not preparing for a headline—we were preparing for an unknown.

And the timeline changed everything. What was supposed to be a standard mission turned into something else entirely. Instead of the planned duration, it stretched into four days. That alone told us—something had gone very wrong. But our job did not change. We just had to be ready… for whatever came out of that capsule."

Q: When the capsule hit the water, what were your immediate orders, and how did the operation unfold in real time?

A: *"I was sitting in the command helicopter, and I remember being absolutely astounded by the re-entry itself. People do not realize how exact that had to be. It had to come in at a perfect angle. Too steep, and they burn up. Too shallow, and they skip off the atmosphere and are gone into space forever. And yet Lovell brought them dead on. They landed on the bullseye. Perfect. I remember just being amazed, thinking, how could they navigate that so precisely? But they did. They put that capsule exactly where it needed to be.*

From there, everything moved fast. Pete was already in the water. I was the last man to go in. Once I got there, I opened the hatch on Apollo 13. That was the unfolding of it in real time—from watching one of the most precise re-entries imaginable, to getting into the water and doing the job we were there to do."

Q: What do you remember about Pete Carolan specifically during Apollo 13—his role, his demeanor, and how he performed under pressure?

A: *"I do not remember specifics about any one individual guy out there. We did not operate that way. We operated as a team. Everyone knew their role, and everyone executed. That is what mattered.*

But I do remember Pete... just not in the way you might expect. The morning after the astronauts were picked up, I saw 'Freddy the Frog.' And that stuck with me. Pete's artwork—he was always creating. Even back then, he was constantly working on military artwork. That is how he has stayed in my mind over the years. Not tied to one specific moment in the water, but through what he created. I still have one of his Apollo 13 posters, the one with all the divers. That piece stayed with me, and I have chosen it to begin my chapter here in this book."

Q: In the water, there is no margin for error—what level of trust existed between you and men like Pete Carolan, Al Starr, and the rest of UDT-13?

A: *"In training, it starts there. BUD/S is so hard to complete. Right from the start, every time you are in the water, you have a swim buddy. You are responsible for each other. I remember being paired up with a weak swimmer. And that teaches you something right away—you do not get to choose. You adapt, and you make sure both of you get through it. That is where it begins. By the time you graduate—by the time you have made it through that level of training—the trust is already there. It has been built. You do not have to* *think about it. So, when you are in the water on a mission like Apollo 13, that trust is not something you question. It comes from what you went through together long before that moment."*

Q: You were trained to stay mission-focused—but was there a moment, even briefly, where the magnitude of what you were doing hit you?

A: *"No. This was my mission, and I had to complete it. That is how I looked at it. I saw it as a great adventure. I never really stopped to think about its magnitude the way people might expect. Even the bigger question—why we were going to the moon in the first place—we never asked.*

You would hear what President Kennedy said, and everything around it, but I never heard what I would call a great explanation. We just did the job. What I do remember... is the capsule itself. It was covered in gold foil, and once it hit the water, that foil started peeling off. We were literally stuffing pieces of it into our wetsuits."

Q: The helicopter artwork— 'The Painted Frog'— "Freddy the Frog," has become iconic. From your perspective, what did that symbol represent to the men on that mission?

A: *"It meant—we were there. We did it. That image... It was one of the first things the astronauts saw when they got* *onto the deck. One of the first visuals was after everything they had just been through. And that stayed with me. Pete left a mark. That frog—that was his mark."*

Q: The world celebrated the astronauts—but your team made their return possible. How do you reflect on the role of the frogmen now, decades later?

A: *"Two years ago, I went to Normandy for the 80th anniversary of the invasion. We were there to dedicate a monument to the Scouts and Raiders, the predecessors to the UDTs and SEALS. Those men cleared the way at places like Omaha Beach. They took enormous losses—about a 50 percent casualty rate. That is where this all started. There is a monument there now, dedicated to them. I remember standing on that beach with a group of us—old frogmen—alongside active-duty SEALS. We took a photo together right there on the sand.*

It meant something to me. I have always been somewhat of a historian. My father was part of that Normandy invasion—he was on a ship and went ashore the day after. So, for me, it is personal. When you look at it over time, there is a clear line: Scouts and Raiders... then UDTs... and now the SEAL Teams. That is the legacy. That is the continuum we are part of."

Q: If you had to define the men of UDT-13—their character, their code, their legacy—what would you say, and where does Pete Carolan fit within that?

A: *"He was a team player. That is the first thing. Whether it was a platoon or a smaller team—it did not matter. That is what defined the men of UDT-13. Even though my own military career was relatively short, it was the most remarkable time of my life. The men I worked with, the men I trained with—they shaped me for the rest of my life.*

When people talk about UDT or SEAL training now, it has become almost legendary. And people always ask—why. I always say that it is the camaraderie. It is a shared experience. And it all starts with BUD/S. That is the common bond. When you meet someone else who has been through it, there is only one question that matters: 'Were you in the Teams?' Because if you were, you already understand.

What it takes is not what people think. It is not about being the biggest or the strongest. It is not about being a 'Rambo.' It is about being mentally tough—and being able to function as part of a team. There was even a study—Dick Couch talked about it—that the typical SEAL was not some massive guy. Generally, men who are five-ten, 160 pounds make it through. The best athletes were not always the fastest or the strongest. Wrestling was one of the best backgrounds you could have. Because what mattered most… was adaptability. That is what defined the men. And Pete, well, he fit right into that."

PATRIOT SEAL HERITAGE FOUNDATION

THE PIRATE & THE PAINT

LANCE GREEN: BUD/S CLASS 42
PSHF BOARD MEMBER

DEANNA GREEN,
PSHF BOARD MEMBER

We first met the man we came to call "Pete the Pirate" the same way so many meaningful relationships in our lives began through the quiet, unspoken bond of brotherhood. My husband, Lance Green, BUD/S Class '42, and I started attending the West Coast Navy UDT/SEAL reunions in 1992. At the time, we thought we were simply reconnecting with old teammates, revisiting shared history. What we found instead were people who would become permanent fixtures in our lives who did not just pass through but rooted themselves into our story. Pete was one of them.

Our daughters were just toddlers then, two and four years old, still seeing the world through wide eyes and imagination. To them, Pete was not just a man, he was a giant. Six-foot-five, larger than life in every sense, with a presence that filled the room before he even spoke. But what they remember most was not his size—it was his warmth. He would scoop one girl up in each arm, effortlessly, like they weighed nothing at all, and his face would break into that radiant, unmistakable smile. It was the kind of smile that made everything feel safe. Pete did not just visit our family. He chose us. And we chose him right back.

Pete's life lived just slightly outside the lines, like he was chasing something just beyond his reach, but enjoying every second of the pursuit. He owned this beautiful wooden boat off the coast of San Diego, a classic vessel that carried a kind of quiet elegance named *"Patriot Seal."* He used to tell us it was the twin to Elizabeth Taylor's yacht, and knowing Pete, you never quite knew where the line between truth and storytelling blurred—but it did not matter. The story felt real because he believed it.

Of course, the boat had its own personality too. There was a large hole in the galley floor, something most people would consider a problem. Pete treated it like a detail.

Because he often operated as more of a "guest" than a paying slip holder, getting to the boat became its own adventure. We would sneak through the gates, timing it exactly right, slipping past the formality of marina rules. But the moment we stepped onto that teak deck, none of that mattered. Out there on the water, everything shifted.

By then, our girls were teenagers, stretched between childhood and adulthood, and those days cruising through San Diego Bay became something unforgettable. A pirate flag flew overhead, snapping in the wind, and there we were—our daughters sunbathing on the bow, laughing, alive in that golden California light. Other boaters would glance over, curious, even a little envious. Pete loved that. He stood taller in those moments, proud in a way that did not need words.

Every third week in August, the Navy SEAL reunion would bring us all back together again, like a rhythm none of us questioned. As the years passed and our daughters grew, Pete began introducing us to another side of his world—his "gallery," Danny's Bar.

To anyone else, it was just a bar tucked away down an alley. To Pete, it was a canvas. He was an airbrush artist of incredible talent, though "business model" was never something he fully embraced. There were times he had been eighty-six'd for an unpaid tab, but that never stopped him. His art lived on the walls themselves—colorful, alive, impossible to ignore.

We would walk down that alley together, sometimes distracting the bartenders just long enough to sneak Pete in through the back. Not so he could drink—but so he could see his work again. He would stand there, taking it in, like he was reconnecting with a part of himself he had left behind. There was joy in him in those moments, a quiet pride as he showed us each piece.

At McP's Irish Pub in Coronado, a famous watering hole for Navy SEALs, I still find myself pointing out his artwork whenever we are there. It is like saying, *He was here. He mattered.* But Pete's life was not just color and laughter. Beneath all of it, he carried something heavier—the weight of his service.

My husband, Lance, understood that weight he carried in a way few others could. He had fought his own three-year battle through the VA system for PTSD and injuries, eventually securing 100% disability. It was not just a process—it was a fight. So, when Pete came to him and said, *"I want to get my benefits,"* Lance did not sugarcoat it for him. *"You're going to have to tell the stories, Pete,"* he told him. *"You are going to have to admit to having PTSD. You will need to see a psychologist, a psychiatrist. It is not easy."* And it was not.

For years, they spoke every week. Sometimes more. There were days Pete would call, frustrated beyond words, overwhelmed by bureaucracy, by operators who did not understand him, by systems that did not speak his language. He would boil over, his temper rising fast, and Lance would sit with him through it—calming him down, grounding him. *"I told you it'd be tough,"* Lance would remind him. *"But you can do this."* And he did.

The day Pete finally received his 100% rating was not

just a victory, it was validation. For everything he had carried, everything he had endured. It was a moment that had been earned a hundred times over.

Our home in Las Vegas became something of a gathering place over the years, especially during SHOT Show. Old warriors would pass through, and with them came stories, laughter, and a kind of understanding that did not need explanation. Barry Enoch, author of *Teammates, SEALs at War,* stayed with us and asked for help selling his book. Pete and Barry even picked out my outfit for the short skirt and heels they thought would help draw attention. We laughed, we played along, and by the end of it, we had sold twenty books. Barry was thrilled.

That night, we sat out on the deck sharing a simple spaghetti dinner. These men—Barry, Lance, Pete—they were gentlemen in the truest sense. They did not curse. They did not glorify the darkness of war. They shared the lighter stories, the ones that could be told.

But when I left the room, I could hear the shift. The quieter conversations. The ones that carried weight. It was good for them, I think—to speak those memories aloud, with people who understood why sleep did not always come easy.

It was during one of those visits that Pete looked at our ten-foot white cinderblock wall and saw something entirely different. *"I need to paint your wall,"* he said. A few years later, he showed up with a horse trailer full of paint. He and Lance went out and chose car paint, of course they did, and Pete got to work. He wanted to paint Diamond Head, but I stood firm. *"I don't live in Hawaii, Pete,"* I told him. *"I want the sea and the sky."*

For three straight days, he worked without stopping. From morning through long after dark, wearing his white overalls, completely immersed in the process. He brought that wall to life, transforming it into something breathtaking. And then, as an afterthought, he airbrushed a tiny ceramic frog nearby. He signed it with a flourish. A piece of his soul, right there in the Nevada desert.

With his VA money, Pete bought his dream car, a 2002 V8 pale yellow Thunderbird. He called one day and said he was driving in from Texas and would stop by. When he pulled up, we could not help but laugh. This towering 6-foot-5 man climbing out of that low-slung car—it looked almost painful. But he loved it, absolutely loved it. He was still our starving artist. But now, he had something tangible to show for it. That car still survives him today, running in the capable hands of my fellow PSHF board member and friend, John Bennett.

Pete continued leaving his mark everywhere. He airbrushed the wheel cover of Lance's '98 Chevy van with his iconic "Freddy the Frog" and a SEAL. It is still there—weathered now, a little beat up—but still unmistakably his. Pete lived in extremes. High highs with low lows. At our 30th anniversary "Mob" party in 2014, he burst in with a bullhorn shouting, *"This is Eliot Ness! You are all under arrest!"* Everyone froze—fedoras, trench coats, flapper dresses—until they realized it was Pete. I walked up, tugged his sleeve, and said, *"Mr. Ness, Prohibition is over. Let's have a drink."* He stayed the night but by morning, he was gone. But he left behind a hand sketched drawing of a frog and a frog bride in 1930s clothes for us.

Pete's temper, though, could turn just as quickly. One night he called Lance from the Tropicana after knocking out a security guard at the Flamingo and escaping through a window. He stayed with us for a few days after that, always looking over his shoulder. But even then, he came back to his art. Sitting at our table, sketching grapes, and wine bottles on a napkin. I offered him paper. *"No,"* he said. *"Most of my best ideas are on bar napkins."*

That was Pete. No structure. No concern for money. Just the need to create. We tried to help him. Suggested canvases, street sales, anything that might give him stability. *"I don't want to cheapen my art,"* he told us. So, we gave him gas money instead and sent him on his way back to California, back to whatever came next.

In his later years, Pete found PeggyAnn Morley—a high school friend who finally brought some structure to his work. She began selling his art at reunions, and we all thought the same thing: *Finally*. Someone who could help him get the recognition he deserved. He would still give his work away, donate paintings, and hand out posters. That was who he was. But with PeggyAnn there, there was hope that Pete's talent would finally be valued the way it should have been all along.

And then came news we were all unprepared to receive in August 2022 at the Navy SEAL yearly reunion. The same place we had seen him every year for decades. But this time, the news was different. Pete was in the hospital. We asked Peggyann every day how he was doing. And then, that weekend, he was gone. The loss of our dear friend who felt like family was not just emotionality painful, it was physical. A deep, aching absence. We realized we would never again hear him call out "Twins!" as he hugged our daughters, now grown. Never see that bone-crushing handshake between him and Lance, followed by that bear hug that lifted them both off the ground. Pete had been a pirate. A warrior. A starving artist. But to us, he was our family.

Today, Lance and I carry that legacy forward as board members of the Patriot SEAL Heritage Foundation.

Along with Peggyann who sits as the President of the foundation, the work we do is rooted in everything Pete's life represented—the visible brilliance and the invisible weight carried by those who served their country.

We are building programs that meet veterans and first responders where they are, not where the system expects them to be. Through K9 training initiatives, we help create bonds that restore trust, regulate the nervous system, and provide a constant companion for those navigating PTSD and TBI. There is something powerful about that connection, something that does not require explanation.

We also support three-day equine retreats, where veterans and first responders step away from the noise and into a space where healing can begin. Horses respond to energy, not words, and in that exchange, something shifts. Walls come down. Regulation returns. For many, it is the first time they feel grounded again.

These programs address PTSD, traumatic brain injury, and related conditions in a way that honors the individual, not just the diagnosis. Because healing is not clinical alone, it is human. And when Lance and I think about Pete, we think about that wall in our backyard. The one he transformed. The one he saw before anyone else did. That is what this work is for our family. Seeing what is still there, even when it may be buried, and bringing it back to life.

GHOST RIDERS IN THE SKY

JOHN BENNETT
VP OF STRATEGIC PARTNERSHIPS
LOS CABALLEROS

I realize the profound responsibility to carry on Pete 'the Pirate's' memory and Legacy as a member of his foundation. I am comforted knowing I will never walk alone in this endeavor. It is not lost to me knowing *"The Many"* out there, with personal bonds, that will not be heard in this work for many different reasons. Telling these stories in remembrance is needed to keep legacy alive and to honor those who came before us. To those who stay behind the scenes due to personal preference, active duty, or other reasons, my aim is to never speak for you but to honor all of you and our shared fortune to have known 'the Pirate' in a personal and special way.

To use our knowledge and stories to continue his legacy in the future through the foundation of PSHF is an honor. One of Pete's most important desires and missions of the foundation when he founded it was to continue to honor and promote service now and in the future. The focus will always be about the TEAMS, brotherhood, and collective good for our military veterans and first responders.

I met Pete some years back at one of the many gatherings of the Rancho Alamitos High school Class of 1965. My mother, Peggyann Morley, and Pete went to high school and graduated together. I was told I must see how close this class still is and meet all the interesting people. I said sure, but thought this would probably be awkward as I would be the youngest there by over 30 years and, what could be so interesting about a high school class? I was blown away immediately by this group and made some good friendships with several people from that first night. From that moment forward, the class always asked my mom to, *"bring John to the event."*

After the first party, I only met Pete quickly and exchanged a few quick pleasantries. On the next visit, we really hit it off. We talked all night, and I helped him set up a brand-new propane outdoor heater he bought from Costco. Going into Costco was a constant problem for Pete because no one on earth had a clue what he might come out with. A pirate ship with lights, another Christmas tree (and it wasn't Christmas), a skull and bones for Halloween (also not around Halloween), or just random clothes and a propane outdoor heater. That night was the beginning of a colorful and impactful friendship.

We exchanged numbers and I would continue to see him at class parties. Many were held at his significant other and President of Patriot Seal Heritage Foundation, Peggy's house. We would also meet out for drinks, and he would invite me down to his boat in Mexico, The Patriot SEAL. He would call me blasting The Beach Boys in the background and just yell *"Arrrrghhh"* into the phone while his dog Blondie would howl in turn after every pirate call.

Eventually, at a Halloween party at Peggy's house, we were talking about his time in the TEAMS and camping outdoors throughout our lives, and he mentioned a horseback riding group he was a part of, and he wanted me to attend a ride. He would be heading off around Saint Patrick's Day. I said, *"Sure, that sounds great!"* Finally, the dates were approaching, and Pete, being Pete, never really gave me much detail or insight into what to expect.

When I asked what I needed to bring to the camping trip, Pete's exact words were *"nothing, I got a bitchin' 10 X 10 tent I'm bringing, so it's all taken care of. You just need to bring some cowboy boots."* So, I got to his house at about 4:30 in the morning as planned. We went to pick up the bar and trailer to hitch it up to the ranch in the Santa Barbara area. When we arrived, he had me set up a huge American flag, draped some camo netting off his SUV trunk for shade, it was extremely hot, and pulled out a lounge chair with a cushion to lay on under the netting. He then pulled out a piece of 4 X 6 astro-turf, threw it on the ground, and said, *"that's for you and sorry I think the cat pissed on it."*

That was the welcome to my first ride with Pete. I knew nothing about the logistics and was not even told to bring a cowboy hat. None of that mattered, and we rode, sat around camp, talked, and had a blast. He never knew this, but before we left, Peggy asked me, *"John, please keep an eye on him as he has had some medical issues come up."* So, I never left his side and never wanted to. It was an amazing experience, and our deep bond was formed along those riding trails.

Time passed and we stayed close. Continued calls, occasional beverages, and lots of laughs ensued. Then we got together again in the early summer, and Pete told me he wanted to sponsor me for membership in Los Caballeros. He informed me I would have to be a gunsel, and with a wry smile said, *"you are going to be bringing me buffalo milks at 5:00 AM."* I clearly had no idea what he was talking about or what I was getting myself into.

I remember going home that night and googling the word "gunsel." That did not provide any more knowledge or insight. That was common with Pete. If you followed Pete, it was generally with little knowledge of what you were getting into. It was just the way he went through life. He eventually just told me, *"Be ready Brother John, I'm going to talk to everyone"* and *"fill out a packet that comes to you, it's important."* I followed his advice as instructed.

Later that summer he called me so filled with excitement that I could feel it through the phone. Pete was always passionate, but I had never heard him like this. He told me, *"I just got the word- you are in. You are going to Catalina with me on the ride."* I was extremely excited as well but still had no idea what I was becoming a part of. I said, *"Great, when is the ride?"* He told me about the dates in September, and I checked and said, *"Pete, I have tickets to the Airshow at Miramar that weekend."* Looking back now, I know how funny that comment is. I also understand why one of the biggest ass-chewings possible would come next. Pete started in and said, *"Air show? F**K an air show! Do you know how many air shows I have been to? Do you know how many jumps I have? You do not have a damn clue how big of a deal this is!"* That rant continued for a few minutes and I responded simply, *"Roger that, F*** the Airshow!"* He was so right; I had no idea what a big deal it was and what an honor it is to be a member of the Los Caballeros.

As the summer leading up to my first official Catalina Caballeros ride went on, Pete and I continued our brotherly routine and still no info on the ride. If I asked, he would say *"don't worry about it."* He did have one focus related to my first ride. He found out I had a connection to the USC marching band through my mother and a Board of Directors committee. He told me he wanted me to get a small 10-man quartet of the band to come to the Queen Mary, where we all leave for Catalina and have them play *Ghost Riders in the Sky* by Johnny Cash. This was Pete's favorite song.

He was relentless with this request and task. So, I went to work to make it happen. I followed up and contacted the right people, and it would cost some money and food, but it was doable. I was then told I needed to submit it to the board. I was thinking, I am nobody in this group, no one even knows me. Nevertheless, I made the submission and calls and was told *"we'll look into it."* I was also asked *"did I know not everyone on the ride went to USC."* I then called Pete and said it did not sound promising and told him the situation. He told me, *"Listen Brother John, don't worry, just get it done. If there is one thing I learned in the TEAMS, it's to never ask for permission, always ask for forgiveness."* I laughed and told him OK, but it was clear it was axed from above.

The summer continued, and everything seemed normal – I had about a week's vacation planned years in advance to Alaska with my family. When I left, I called Pete and told him I would be back soon, and it all seemed well. I returned home, was back only one day, and received a call from Peggyann that Pete was in the hospital. I called at once and he sounded okay and was full of piss and vinegar. He was mad that the nurses would not give him the ice cream and the drinks he wanted. He told me he needed me to go on a mission on my way to see him. He said, *"I need you to smuggle in some snacks, ginger ale and ice pops."* I said, *"I'll get on it and be there soon."*

When I got to the hospital, I showed up in full cowboy gear for the first time. He lit up like a Christmas tree, he was smiling from ear to ear, and called Peggyann immediately. He enthusiastically yelled through the phone and said, *"Peggy, a real-life cowboy just showed up in my room!"* Pete had made the room his own with a SEAL Team recruitment poster, one that he had painted, hanging on the wall. And for better and worse, the nurses knew him well. He looked good, was full of life, and was excited to be getting out soon.

Pete was excited to ride in Catalina Island , excited to go to a few USC Trojan football games, excited to go to the Famous 1923 Club at the Coliseum, and eat lobster rolls. I did not want to break it to him that the food was nothing to be excited about. It has since improved greatly and sure enough this last year added a lobster roll to the menu. When I first saw that I laughed, looked up, and thought *"only Pete."* I went home from the hospital that night in great spirits, expecting to see my friend soon, and for us to head out to Catalina together.

I got the call early the next morning, my friend and my brother was gone. Like all who eventually heard the news of Pete passing away, I was in shock and disbelief set in. As word started getting around and we all began to meet to gather and celebrate Pete's life, it was apparent some found out that I had seen him recently in the hospital, in fact I was the last person to visit Pete. Pete never believed his illness was that serious, he was a tough Frogman, so he figured there was no reason to call anyone and that he would be out soon. I saw him personally and I believed the same.

Pete was larger than life, with such magnetism; no one believed it was possible with a person like him to just be gone so suddenly. As we all gathered, I received questions about his last days, and I almost sensed a feeling of jealousy or envy that I was privileged to have those precious last moments that they wished they could have had with Pete – and I understood the feeling.

Pete, while never constrained by it, seemed to have a space and place for everyone. To know Pete personally, it was a connection that only that person could understand but was truly whole. As if for a moment in time, it would seem the only relationship that existed. All realized that these moments were fleeting. Pete inevitably exited quietly to the next journey precisely opposite the loud and large way he had just entered . That was the pleasure and the bounty of being close to the Pirate.

The next few months for me were foggy and consisted of many gatherings, stories, several celebrations of life, and trips up and down the California coast. Once or twice in his prized two-seater T-Bird coupe he left behind. Pete lay to rest in Miramar National Cemetery. He also has a dedicated brick at the VFW in Coronado where pieces of his military combat art can still be found hanging on the walls.

Around this time, the ride with the Los Caballeros was quickly approaching, and Jim Cunningham took me under his wing a bit when it came to riding. I rode horses a bit as a kid growing up, but until the St. Patrick's Day ride with Pete, my last time on a horse was a 45-minute ride at age 17 that could have been a scene in City-Slickers. But these were real Cowboys who did not just ride horses; they lived the cowboy life.

Jim introduced me to one hell of a rider, cowgirl, and person, TJ. TJ is a 100-pound soaking wet Marine, Cobra Helicopter pilot, and absolute badass that can ride the hell out of any horse. I am fairly sure she took pity on me in the beginning and gave me a crash course in riding and one hell of an Arabian horse for the ride on the Island to boot. I am certain that horse saved my ass on that Island and in the rodeo.

Pete and I had that in common and are of the very few that did not come directly from a ranching or complete horsemanship background in the Los Caballeros. Ours was a path forged by different experiences, but those experiences and crossovers are surprisingly similar and valuable. The desire to learn, excel, and never quit; the rest came naturally. We are never short of those willing to spend time and teach those willing to learn and put in the effort.

I would leave for my first-year ride with Los Caballeros on Catalina Island without my brother and sponsor, Pete Carolan. This was a tough time, but I immediately felt everyone around me, ready to lift me up. No one was trying to replace Pete or watch over me, just supportive with small gestures in every way. One thing that remained consistent – I still had no idea what to expect on this five-day ride into the wilderness and back country of Catalina Island. I had no sponsor to tell me anything and missed all the pre-ride gatherings while attending ceremonies.

I showed up on the island with a sleeping bag, floor mat, pillow, and a tarp. The Presidente loaned me an extra cot to sleep on. Jim Cunningham helped me get riding and rodeo lessons. Veterans gave me quick words on small things, such as which hat to wear. David Kirby called the day before to give me a ride over to the boat. Then, my journey and ride were on me. It was open for me to take from it what I will. Freedom that was made possible with just the right number of guiding hands. I will remain forever grateful for that experience.

After a blur of an amazing week that flew by, our Caballeros Cowboy brothers held our annual empty saddles ceremony. This is a special cowboy tradition for all brothers lost that year. This was followed by reading aloud every member name of the history of the club that now rides the dusty trails in the sky.

On the last morning, a famous Cowboy, Marine, and recording artist, Rusty Richards, at an advanced age, climbed up a steep rocky hill with guitar in tow, up to Cowboy Church. It sits on a bluff, overlooking the cove and harbor that Pete painted many a picture of. One, an original specifically for Rusty of that cove that hangs in Rusty's family home. Rusty strummed that guitar, and we all belted *Ghost Riders in the Sky* among others repeatedly. I was in awe of the heart, dedication, and character that surrounded me that day. It was something one can never forget and must feel a responsibility to pay forward and pay back in fold.

I did not understand exactly until years later with reflection, experience, and hindsight; for an artist like Pete Carolan, the Los Caballeros were a canvas for brotherhood, cowboy camaraderie, and art. It exercised the strength and fragility of the body and mind in a healthy, productive way. It was yet another airbrushed outlet in a different form.

Pete lived the dichotomy of simultaneously belonging to everyone and no one at the same time. To know Pete was to have him come in and out of your life on his time. Such was the way he approached and moved through life. A free spirit, organic in nature like a wave, forming, rolling, building, sometimes crashing, then rolling back out to sea, all to at some point, do it all over again.

Yes, he resembled the one thing he most mastered in life – the ocean. Its own rhythm, its unseen undercurrents, surface beauty, unpredictable nature, fierceness, and comforting profound moments. This is my friend and brother in life experience, and my companion in mission and legacy.

After meeting Pete's family in the TEAMS and attending Pete's SEAL Team swim out at the West Coast Navy SEAL reunion in Coronado, and The Cowboys Caballeros empty saddles ceremony, one thing became clear, there are many similarities that I see Pete was drawn to. Honor being one of them. And a simple, important rule followed; Leave it better than you found it- That applies to the land, the community, and the people around you.

Gratitude has been the quiet undercurrent running through every chapter of this journey. Gratitude for those who raised me from a young age and shaped the foundation of who I became, and for life's twists and turns that refused to let me walk a straight, predictable path. Gratitude for meeting Pete, and for the experiences forged through both the military and cowboy communities—worlds that taught me discipline, resilience, and respect for something larger than myself. I am grateful for my friends, family, and the unwavering support structure that held steady through the storms, and for the team and dedicated unit within PSHF whose commitment gives real weight and meaning to the mission we have been entrusted with.

There is gratitude in being part of something bigger than oneself, in doing the essential, often unseen work of helping our heroes heal, and in witnessing the ripple effects that healing creates across families, communities, and within our own lives. Above all, I am grateful for the relationships I have made, continue to foster, and still find ahead of me through this foundation and its mission—a community of dedicated, talented, full-hearted people who show up for one another without hesitation. It is, in every sense, the best of America.

A PIRATE'S HEART

APRIL EGBERT THOMPSON, VP OF MARKETING, PATRIOT SEAL HERITAGE FOUNDATION

I first met Pete Carolan in January of 2022 through Peggy Ann Morley, a longtime client and dear friend of more than 25 years. Peggy Ann told me, *"I want you to meet Pete. He has a foundation, and I think you would be a great asset."* Pete's initial response was classic Pete: *"I don't think we need another woman on the board."*

Peggy Ann, who serves as President of the Board for the Patriot SEAL Heritage Foundation, arranged for us to meet at Jade Restaurant. Over lunch, I shared stories about my upbringing — how my father taught me to throw a football, shoot guns, deep-sea fish, and hunt, while my mother taught me grace, etiquette, and strength. I explained that although I stood 5'11", I was still a tomboy at heart, equally comfortable at a black-tie gala or filleting fish after a long day on the water.

By the end of that meeting, Pete's perspective had completely changed. What began with skepticism evolved into one of the most meaningful friendships and mentorships of my life.

As a board member of the foundation, I have the privilege of witnessing firsthand Pete's unwavering dedication to honoring the legacy of the Navy SEAL community, preserving the brotherhood, and creating pathways of healing for Navy SEALs and United States veterans, first responders struggling with PTSD, TBI, and related traumatic brain injuries.

Pete was larger than life — a storyteller, super patriot, protector, artist, and loyal friend whose presence could command a room while making every person feel seen, heard, and valued. But beyond his humor and larger-than-life personality was a man deeply committed to helping veterans rediscover hope and purpose after war with his art.

One of the missions closest to Pete's heart — and now a growing reality through the foundation — is the development of three-day PTSD and TBI equine retreats on Catalina Island for Navy SEALs and veterans. These equine retreats, scheduled to begin in the fall of 2026, built through the strength of the SEAL brotherhood, his brotherhood with Los Caballeros, community support, and strategic partnerships united around a common mission: healing our warriors.

These equine retreats, designed for Navy SEALS to

become sacred spaces for restoration — places where Navy Seals carrying invisible wounds can reconnect with themselves, nature, horses, and the brotherhood they often miss after service. Pete understood that healing does not always happen inside hospitals or therapy rooms. Sometimes healing begins quietly around a campfire, on horseback, or in honest conversation with another warrior who understands.

Pete also worked closely alongside fellow board members and Navy SEAL Team One, Andy McTigue, through VETCIA, supporting K9 training programs for veterans and first responders. Andy and his team have championed innovative approaches to healing by pairing highly trained service dogs with those battling trauma, anxiety, and the long-lasting effects of service.

What many people may not realize is that Pete's artwork became an extension of his mission. Through his paintings and artistic expression, there was always an undercurrent of resilience, redemption, loss, and hope. A time capsule of work, his art carried emotion that words often could not express. From Vietnam to 911. For many struggling veterans, Pete's work became a reminder that beauty could still emerge from pain and that purpose could still be found after darkness and that no man will be left behind or forgotten.

During this same season of life, my mother became ill and passed away on July 2. I organized a black-tie Celebration of Life for her at the Balboa Bay Resort and asked Peggy Ann if Pete might attend. She laughed and said, *"Pete hates black-tie events — especially tuxedos."* To my surprise, Pete not only attended but bought a brand-new tuxedo specifically for the occasion. That gesture alone spoke volumes about the depth of his loyalty and heart.

That evening, Pete told me that I was now his adopted daughter and that my boys were his adopted grandsons. It was one of the greatest honors of my life.

Shortly afterward, I attended my first Navy SEAL Reunion in Coronado, CA, representing the foundation, expecting Pete to join us. Sadly, his health had declined, and he was unable to attend. After the reunion, I visited him at Hoag Hospital and brought him a framed photo of my mother's celebration of life. He proudly showed the nurses the picture, introducing us as his adopted family.

Only two months after my mother passed, Pete passed away as well.

Yet Pete's legacy continues to grow through the work of the foundation with the developing equine healing retreats in Catalina, the K9 initiatives supporting veterans and first responders, and through the powerful combat murals and artwork he left behind.

For those who are struggling, Pete "The Pirate" Carolan's life reminds us that even through grief, trauma, and loss, there is still hope for healing and a brighter future. His spirit lives in every Navy SEAL, veteran helped, every partnership formed, every brother who steps forward to serve another, and every piece of art that continues to speak to the resilience of the human soul. For me personally, Pete 'the Pirate' Carolan was far more than a board founder or mentor. He became my chosen family. His spirit continues to guide all of us, I feel fortunate enough to have known and loved him.

FIRST SIGHT OF THE PIRATE

CAROL MARSHALL CRANE, BOARD HISTORIAN, PATRIOT SEAL HERITAGE FOUNDATION

Before I ever met Pete Carolan, I felt as though I already knew him from stories my husband told me. Skip Crane, my husband and retired Navy SEAL Commander, regaled me with vivid, strange tales about his teammate from BUD/S Class 47 back in 1968 in Coronado, CA. Each story ended in at least a few chuckles. Pete left stories everywhere he went. By the time I met him, I was curious if anyone could match such a reputation.

The meeting took place at Coasterra Restaurant on Harbor Island in San Diego, CA, sometime in the late 2000s. As Skip and I waited along the waterfront, he pointed toward a weathered wooden Chris-Craft floating offshore. *"That's Pete's boat,"* he said. Pete's vessel was named *Patriot SEAL.*

Even from a distance, I could tell the boat had seen better days. The once-beautiful wooden hull bore the marks of age and hard use. It looked less like a pleasure craft and more like a faithful companion that had survived countless adventures, including being chased by the Harbor Police a few times. A pirate flag fluttered from the boat, faded and worn from years of sun, salt, and high winds. It seemed perfectly appropriate. If the boat worked, Pete was happy. If there was ever a man who embraced the spirit of adventure, it was Pete Carolan.

The boat sat anchored two hundred yards offshore. As we watched, Pete emerged from the cabin. Pete was wearing nothing but swim trunks and carrying a plastic bag. There was no towel. No concern about the chilly water. No hesitation. Without a second thought, he dove overboard. Skip and I stood there watching as Pete began swimming toward shore dragging his plastic bag through the chilly waters of San Diego Harbor. Seeing his teammate swim, Skip remembered that it was Pete who helped him learn more efficient Navy SEAL swimming techniques. For Pete, it was not a leisurely swim. It was simply his chosen mode of transportation.

Minutes later, Pete reached land, disappeared behind a nearby bush, and somehow transformed himself from harbor swimmer to lunch guest. Moments afterward, he emerged in dry clothes, his hair still dripping water, looking completely unconcerned with appearances. As he approached, Skip introduced us. Pete greeted me warmly, then at once turned to his old teammate with a grin spread across his face. *"That was a great swim,"* he said. *"Let's eat."* It was the most Pete Carolan introduction imaginable.

Inside the restaurant, lunch stretched into an afternoon of stories, laughter, memories and perhaps PBR beers. Pete and Skip retreated effortlessly back into the bond they had forged decades earlier during Basic Underwater Demolition/SEAL training and through their years of service. Listening to them reminisce, I began to understand that their friendship represented something deeper than shared experiences. It was part of a brotherhood formed through hardships, sacrifice, mutual trust, and loyalty. Hooyah!

What struck me most about Pete that day was his authenticity. There was no performance. No effort to impress anyone. The stories I had heard about him were true because Pete simply lived life on his own terms. Whether swimming ashore from an anchored boat, creating artwork, supporting veterans, or sharing stories with old friends, he approached life with enthusiasm, humor, and an unmistakable sense of adventure.

So many people knew Pete Carolan as a Navy SEAL, a patriot, an artist, a storyteller, cowboy, and a lifeguard. Some in the SEAL community knew him as "The Pirate." Others knew him as the man behind *Patriot SEAL.* But on that afternoon in San Diego, I learned something more important, I met a man who remained completely himself.

That first encounter revealed so much of what would later become evident to everyone who knew him. Pete had an uncommon freedom of spirit and a little bit of a rogue personality. He was fiercely independent, deeply loyal to his friends, proud of his service, and always ready for the next adventure. The years may have weathered his boat, just as they weather all of us, but they never diminished his enthusiasm for life. Long after the lunch ended, I remembered the image of Pete emerging from the harbor, swimming ashore beneath the California sun as casually as if everyone arrived for lunch that way.

It was my first glimpse of the man whose stories would inspire *The Painted Frog.* And, in many ways, it told me everything I needed to know about Pete Carolan. As Board Historian of the Patriot SEAL Heritage Foundation, I have witnessed firsthand how Pete's influence continues long after his passing. That lunch at Harbor Island was only the beginning of my friendship with Pete Carolan.

Over the years, Pete became a frequent visitor to our home in Imperial Beach, CA, usually in August when the West Coast Navy SEALS would have their annual reunion. Like so many members of the SEAL brotherhood, he never needed an invitation. He simply arrived, usually with a story to tell, a memory to share, or an idea he was excited to bring to life, and sometimes he would bring his dog Blondie, the love of his life.

Today, long after Pete's passing, his presence remains. Throughout our home are pieces of his artwork—paintings, sketches, and reminders of the remarkable creativity that existed alongside the warrior spirit for which he was so well known. Each piece reflects a different side of Pete. Some capture his patriotism. Others reveal his humor. Many tell stories that only Pete could tell, like life living in the shadows. Together, they form a living gallery of a life fully lived. A part of Pete's ashes also remain in an AMMO can in our garage, where teammates share a toast to "Pete the Pirate."

As Board Historian for the Patriot SEAL Heritage Foundation, I have come to appreciate how important it is to preserve the stories of men like Pete Carolan. History is not merely a collection of dates and events. History lives in the memories we share, the lessons we pass forward, and the people who refuse to let those stories disappear, Pete understood that.

He often spoke about legacy—not in terms of personal recognition, but in terms of service. He believed deeply in preserving the heritage of the Underwater Demolition Teams and the Navy SEALs, United States Veterans, and First Responders who came before. He wanted future generations to understand the sacrifices, the difficult struggles that people feel, the camaraderie, and the values that shaped the community he loved so much. That vision became one of the driving forces behind the Patriot SEAL Heritage Foundation.

Today, that mission continues through the work of dedicated volunteers, board members, teammates, family members, and friends who have committed themselves to keeping Pete's dreams alive. Every event, every educational program, every effort to support veterans, first responders, and those struggling with the invisible wounds of service carries a piece of Pete's spirit. This book is part of the PSHF mission.

The Painted Frog is not simply the story of Pete Carolan. It is the story of the people whose lives he touched. It is the story of teammates who served beside him, friends who laughed with him, artists who admired him, veterans who respected him, and family members who loved him. It is a collection of memories woven together by those who knew him best.

Each contributor has become a steward of Pete's legacy. Some provided stories from combat. Others shared tales of friendship, adventure, and laughter. Some remembered the artist. Others remembered the pirate. Still others remembered the mentor, the patriot, and the man who never stopped believing in the importance of serving something greater than himself.

Together, these voices paint a portrait that no single person could create alone. That is why this book matters. For future generations, Pete Carolan will be more than a name etched into the history of the SEAL Teams. Through these pages, he becomes real again—a young lifeguard from Huntington Beach, CA, determined BUD/S candidate, a Vietnam UDT 13 veteran who served two tours, a military combat artist, a sailor, a super patriot, a cowboy, and a friend.

Most importantly, he becomes a reminder that a life devoted to service does not end when a man is gone, it continues in the stories we tell and in the lives we touch. And it continues whenever those of us left behind choose to keep the home fires burning. Today, that is exactly what we are doing for Pete, for his teammates and for the heritage that he cherished, and for the generations who will one day inherit the history and the stories entrusted to us.

THE PIRATE, THE PLATE, AND THE PATH TO HEALING

KRISTOPHER MARSHALL,
VP CREATIVE & TECHNOLOGY;
THE PATRIOT SEAL HERITAGE FOUNDATION

They say you never get a second chance to make a first impression, but when you are sitting across from a man known as "The Pirate," one impression is usually all you need to last a lifetime. My introduction to the legendary Pete Carolan—the heart and soul behind the Patriot SEAL Heritage Foundation (PSHF)—happened over a breakfast that I can only describe as "ambitious."

At the time, I was running on fumes, bleary-eyed from a night of far too little sleep. Across from me sat Pete, a retired Navy SEAL and a long-time friend of my stepfather, Skip Crane. While I struggled to keep a coherent thought, Pete was busy dismantling an extra-large mountain of bacon, eggs, hashbrowns, and toast.

There is something profoundly humbling about watching a legend tackle a breakfast plate with the same tactical precision he applied to his time in the Teams. In that moment, despite my exhaustion, I realized I was not just meeting a business contact; I was meeting a literal breath of life.

I was there to offer my ability in marketing and brand exposure to help highlight his artwork. Because of the deep history between Pete and my stepfather, I felt a calling to offer my services pro-bono. I wanted to help the world see the man behind the "Pirate" moniker—the artist, the warrior, and the healer.

Fast forward time, a little over four years, and while Pete has moved on to calmer waters, his legacy is still my compass. Today, I serve as the Vice President of Creative & Technology for PSHF. It has not always been smooth sailing. Building a foundation from the ground up is a bit like trying to build a ship while you are already at sea. There were moments of intense frustration, long nights of digital heavy lifting, and the occasional feeling that I was rowing the boat with little aid. I have built and designed the website from the ground up, created and curated our social media profiles and posts, and traveled to countless events to ensure Pete's story did not just stay in the past, but fueled the future.

Now, we enter a new, calculated chapter defined by a dedicated internal team and powerful partnerships. Our mission is simple yet massive: to help our heroes heal.

To provide K9 and Equine Therapy and adventurous retreats to those who need it most.

We believe that healing is not a one-size-fits-all solution. For some, it is the cold nose of a service dog or the rhythmic gait of a horse. For others, it is the quiet solitude of an island adventure or the creative release found in art and music. We serve as a beacon of hope for veterans and first responders, offering them the tools to rediscover their own "breath of life".

I am incredibly proud to be with this team. We are not just managing a foundation; we are stewarding a legacy of service. We are here to celebrate the man, the artist, and the "Pirate" who showed us that even after the roughest nights, there is always a reason to pull up a chair, order the extra-large portion of bacon & eggs, and get to work.

A BROTHERHOOD REMEMBERED

TOMMY BENNETT, LOS CABALLEROS MEMBER & PSHF BOARD MEMBER

The legacy of Pete Carolan cannot be told through a single lens. It lives in the stories of fellow Navy SEALs, artists, veterans, family members, and friends whose lives were forever shaped by his presence and brotherhood.

Among those voices is Tommy Bennett, known throughout the Los Caballeros community as "Tommy Gun." What began as a chance meeting on Catalina Island evolved into a friendship built on trust, loyalty, and a shared commitment to preserving the values Pete held dear. Through years of conversations and adventures, Tommy witnessed firsthand the qualities that made Pete larger than life—not simply as a Navy SEAL or artist, but as a man who dedicated himself to building connections, strengthening communities, and honoring the brotherhood he loved.

Today, Tommy serves as a Board Member of the Patriot SEAL Heritage Foundation, helping advance the mission Pete worked so passionately to establish. In this conversation, he reflects on friendship, brotherhood, and the enduring legacy of a man he considers both a hero and a legend.

TELEPHONE INTERVIEW CONDUCTED BY DR. (H.C.) LISA RAY WITH TOMMY BENNETT PSHF BOARD MEMBER. JUNE 3, 2026.

Dr. (h.c.) Lisa Ray: How did you first meet Pete Carolan, and what was your first impression of him?

Tommy Bennett: *I first met Pete during one of our annual Los Caballeros Catalina horseback rides. It's one of those traditions where you end up seeing the same faces year after year, but Pete stood out at once.*

He showed up dressed differently than most of the riders—old school historical mounted cavalry —with a well-worn hat that looked like it already had a few stories behind it. There was something unforced about him right away. No performance. No ego. Just a guy being himself. We connected quickly. He had a natural humor—sharp but never cutting—and a way of making people feel at ease almost immediately. That first impression never really changed. If anything, it became clearer over time.

Dr. (h.c.) Lisa Ray: At what point did your friendship evolve into something deeper than a causal connection, and what made that bond so strong?

Tommy Bennett: *It developed naturally overtime as we spent more of it together. At that point, I was a partner and investor in a waterfront restaurant in Long Beach called Jade, and Pete took a genuine interest in it. Not casual interest—he wanted to understand it, ask questions, and think through how it could grow. We started meeting regularly with other Los Caballeros members, often on Tuesdays. Those meetings became less about socializing and more about real conversation—life, business, purpose, and experience.*

Pete was always thinking creatively. Some ideas were big, some were unconventional, but they all came from a place of wanting to help others succeed. What really deepened the friendship was trust. We began sharing life stories in a more honest way. I was drawn to Pete's journey—from USC to him becoming a Navy SEAL UDT-13, and later an artist and storyteller. The more I learned, the more respect I had for the depth of what he had lived.

Dr. (h.c.) Lisa Ray: During the pandemic, Jade became a gathering place for veterans, friends, and community members. What was happening inside those walls that people may never fully understand?

Tommy Bennett: *The pandemic created a level of uncertainty that touched everyone. Like many businesses, we had to adapt quickly. Jade ran in a quieter, more discreet way during that time, a speakeasy for my closest friends. But what emerged wasn't just about keeping a restaurant open, it became about maintaining connection. People needed a place to go where they could still see familiar faces and have real conversations when so much of the world felt shut down.*

My friends, and community members came through, and there was an unspoken understanding that this was more than just a restaurant at that point. It was one of the few places where people could still feel grounded and connected.

Dr. (h.c.) Lisa Ray: What role did Jade play emotionally and spiritually for Pete during that time?

Tommy Bennett: *It wasn't something we set out to intentionally design, but that's what it became. For Pete, structure and connection are important. During the pandemic, both disrupted. Jade provided consistency. People knew they could show up, see familiar faces, and be around others who* *understood the importance of camaraderie. It wasn't therapy, and it wasn't formal support. It was simply a human connection. And sometimes that's enough to carry people through difficult seasons.*

Dr. (h.c.) Lisa Ray: People know you as "Tommy Gun." Where did that nickname come from?

Tommy Bennett: *{Laughs} It goes back to my beach volleyball days. I had a pretty aggressive spike when I was younger, and could jump. A friend started calling me "Tommy Gun." It stuck for a while, faded out, and then resurfaced again when I joined Los Caballeros, where nicknames are part of the culture. I don't take it too seriously, but it fits in a way.*

Dr. (h.c.) Lisa Ray: Pete gave you one of his Tridents. Can you describe that moment and what it meant to you?

Tommy Bennett: *That moment is one of the greatest honors of my life. Pete shared the story behind the Trident and what it represented in his service and identity as a Navy SEAL. I understood the weight of that immediately. Over the years, we had many conversations about his experiences, and I never took those lightly. I respected everything he had lived and what it meant.*

When he handed me the Trident, it wasn't just symbolic—it felt deeply personal. It was a gesture of trust and friendship. He also gave me a custom name badge and signed the back of it for the Los Caballeros ride.

I attached the Trident to that badge and keep it in my truck to this day. I see it every day and it's a reminder of Pete and the values he lived by.

Dr. (h.c.) Lisa Ray: For those who may not understand the significance of a Navy SEAL Trident, what did receiving that symbol from Pete represent to you?

Tommy Bennett: *I have deep respect for what the Trident stands for, especially knowing I did not earn it through the path Pete did. That is exactly why it meant so much. Receiving it from Pete felt like an expression of trust and respect that goes beyond typical friendship. It stood for loyalty, brotherhood, and shared values. Not through shared service—but through shared integrity.*

Dr. (h.c.) Lisa Ray: As a Board Member of the Patriot SEAL Heritage Foundation, why did you feel called to support the mission?

Tommy Bennett: *My connection to the Foundation began through my friendship with Pete. As I learned more about his vision, I came to understand that the Foundation is about much more than preserving history. It is about honoring* *service, supporting veterans, educating future generations, and ensuring that the SEAL community's legacy is preserved with dignity.*

Pete believed deeply in that mission. Supporting it felt like a natural extension of respect for him and for what he stood. My role is supportive, but I take that responsibility seriously.

Dr. (h.c.) Lisa Ray: In your opinion, what made Pete Carolan different from other men you've known?

Tommy Bennett: *Pete was larger than life, but never in a way that felt forced or performative. He could tell stories that sounded unbelievable until you realized others had been there too. That's when you understood he wasn't exaggerating—he had simply lived an extraordinary life. He had a sense of humor, intensity, creativity, and heart all at once. He could elevate any room he walked into, but he also genuinely cared about people. He was a true American patriot in how he lived, not just how he spoke.*

Dr. (h.c.) Lisa Ray: When people look back on Pete's legacy, what do you hope they remember most?

Tommy Bennett: *I hope they remember his honor. The art, the stories, and the experiences matter—but underneath all of it was a man deeply committed to his brothers, his country, and his values. Pete didn't just speak about patriotism.*

He lived through his service, his friendships, his creativity, and his dedication to preserving legacy. If there is one thing to carry forward, it is that he lived with loyalty, purpose, and integrity. That is what should be remembered about Pete.

FREDDY THE FROG

ARTIST: PETE "THE PIRATE" CAROLAN 1970

FORGED OUTSIDE THE LINES

ANDY MCTIGUE / SEAL TEAM ONE-PRESIDENT OF (VCIA) VETERANS CANINE INTELLIGENCE ACADEMY 501(C)(3)

Board Member & Strategic Partner of Patriot SEAL Heritage Foundation

I did not set out to become a Navy SEAL—I collided with it, the way certain men collide with their destiny, not through planning but through friction, restlessness, and an unwillingness to sit still in a life that does not fit. At twenty years old, I entered the Navy already carrying more life experience than most of the men around me, and that mismatch showed immediately. Aviation electronics school rejected me—not because I lacked intelligence, but because I could not be contained. A psychologist told me plainly: *"You should have been an officer, but you're not, and now you don't fit anywhere."* So, the Navy did what it often does with men it cannot easily categorize—it sent me to the fleet.

When I finally found myself in Subic Bay, waiting for a ship that could not seem to stay in one place long enough for me to board it, boredom pushed me toward a small booth labeled UDT/SEAL screening. It was barely larger than a phone booth, but inside that space was a doorway into something that would define the rest of my life.

I walked in, spoke to the officer, answered a few questions about swimming and athletics, and, within 30 minutes, was told to show up the next morning. That was it. No ceremony. No buildup. Just a simple directive that would separate the rest of my life into before and after. I reported to the USS Ticonderoga and informed my CO that I had made the ACDUTRA unit. He crumpled the paper in front of my face and told me to forget about that. I was his for the next nine months.

The USS Ticonderoga became my classroom, but not the kind with desks and instruction manuals; it was a floating crucible of fire, fuel, and constant proximity to death. Hypervigilance was not a symptom—it became a baseline. Fire was not an emergency—it was daily life. And death was not distant—it was always within reach, hovering just close enough to sharpen the senses without taking me out. I spent nine months on that deck fueling planes, watching planes crash within feet of me, seeing propellers shear into sparks across steel, feeling the heat of fires that seemed inevitable and unstoppable, all while holding a fuel line that could have turned the entire ship into an inferno. The reality is unmistakable: that environment rewired me, and at times, I still struggle with PTSD from those years in service.

When my nine months ended, I reported back to SEAL/UDT. What followed was not just training, it was exposure. Exposure to the limits of the human body, and more importantly, the human mind. "Hell Week" did not break me because I understood something fundamental that many did not: they were not trying to kill me. They were trying to make me believe they could. And that distinction gave me control. I stood there while instructors screamed, conserving energy while they expended theirs, doing exactly what was required and nothing more, understanding that survival in that environment was not about dominance, it was about endurance. Each morning, as the sun rose behind my back and warmed my skin, it marked one more day completed, one more day survived, and that was enough. Not the whole week. Not the whole program. Just one more day.

That is the mindset that separates those who stay from those who ring the bell. It is not grand. It is not heroic. It is quiet, deliberate, and relentless. I graduated from BUD/S, moved into SEAL Team 1, and was sent to Vietnam, carrying with me not just the training but the mental framework that would allow me to operate amid chaos without becoming consumed by it. Years later, in the reserves, that same framework would intersect with a man who operated by a completely different internal compass—Pete "the Pirate" Carolan.

Our meeting was unceremonious, the kind of introduction that happens when men cycle through the same spaces, one weekend a month, two weeks a year, checking in, training, maintaining a connection to something that never quite leaves them. At first, Pete was simply another presence—louder than most, rougher in his delivery, carrying an energy that did not seek approval and did not require it. But over time, it became clear that Pete was not simply different—he was operating on a frequency that did not align with the structures around him.

Within the Teams, difference is not immediately rejected—it is assessed. Pete was unpredictable, yes, but not in a way that threatened his brothers. There was a line, and he did not cross it. He respected the men beside him. He understood the bond. And that alone earned him a place. The outside world, however, was another matter entirely. There is a story I remember from those days—a small almost insignificant on the surface—about a candy bar.

Pete had no money. Not temporarily broke, not between paychecks—he lived in a constant state of having nothing. He walked into a gas station with change in his hand, trying to buy something simple, only to be stopped by a request for identification. At that moment, something inside him ignited. Not because of the transaction, but because of what it represented: a barrier, a lack of recognition, a system that did not see him for who he was. He reacted instantly, throwing the candy bar down, watching it bounces up and strike the woman behind the counter, leaving a mark on her face that turned a minor incident into a disciplinary one. He was ordered to return and apologize, and he did.

That matters because it shows that while Pete could erupt, he was not without accountability—at least within the framework of the brotherhood he respected. This duality defined him. He could be volatile in the civilian world but grounded within the Teams. He could live without money, without stability, without a fixed place to land, but still operate with a deep sense of loyalty to the men he served beside.

And then there was the art he created. It is tempting to separate Pete's art from Pete himself, to view it as a product, a talent, or a skill set that can be analyzed independently, but that would miss the truth entirely. His art was not something he did, it was who he was. I remember watching him build what would become known as the *Painted Frog Float,* and the description is less about technique and more about instinct. There was no formal plan, no structured design, no visible roadmap—just buckets of expanding foam poured into a rough structure, growing into an undefined mass that made no immediate sense to anyone watching. Thirty buckets later, there was still no clarity for anyone except Pete. And then he began to carve. What emerged was not accidental, it was inevitable. A massive frog took shape, ten feet tall, fifteen feet long, painted with intention, personality, and presence. It was absurd and brilliant at the same time; a physical manifestation of the way Pete saw the world. Where others saw a blank wall, he saw a finished piece waiting to be revealed. Where others required structure, he required only space. This ability—to see beyond the surface, to create without permission, to bring something into existence simply because it existed in his mind—is what separated him.

And yet, despite this gift, he never translated it into financial stability. He did not monetize his work. He did not build a career from it. He lived on floors, on borrowed space, on the margins of comfort, often relying on the goodwill of others just to maintain basic existence. I can remember paying fifteen dollars to get Pete into a reunion because Pete did not even have that. That detail, small as it is, reveals the larger truth: Pete was not running on the same metrics as everyone else. He was not building toward security—he was expressing something that did not fit into that framework. And yet, his work found its way into the highest places—the walls of the National Navy UDT-SEAL Museum, the Pentagon, the White House—recognized not because he chased recognition, but because what he created could not be ignored. His pride was not in ownership—it was in association. Being part of the Teams and being connected to that lineage, that was enough.

Pete's identity was inseparable from that connection. His art was not about ego—it was about belonging. Even the emerging conversation around his potential contribution to the design elements of the SEAL Trident reflects this. Not a man claiming credit, but a man whose fingerprints may exist within something far larger than himself, a symbol worn by generations of warriors. And yet, even here, the approach is careful, deliberate, respectful of the institution that holds that history. It is not about rewriting—it is about acknowledging.

Today, I carry that same ethos forward, not through art, but through service. My work with veterans—pairing them with dogs, guiding them through training, connecting them with resources—is not framed as charity. It is framed as a continuation. I am not a psychologist, and I do not claim to be. I am a man, a veteran, who understands what it means to live with the aftereffects of service, to operate in a world that feels slower, softer, and often disconnected from the intensity that once defined daily life.

I meet veterans where they are, not above them, not outside them, but beside them. My foundation helps them build relationships with dogs that are not just companions, but anchors—living, breathing connections that provide structure, responsibility, and in many cases, a reason to stay. Five men have told me directly that without this program, they would not be here, and they chose life over suicide. Five who admitted it. The number beyond that is unknown, but the impact is undeniable. This is not abstract work, it is immediate, personal, and often lifesaving. And in that, there is a direct line back to the men like Pete.

Because while Pete may not have lived a conventional life, he embodied something essential: the refusal to disengage. The refusal to step away from the identity that defined him. He did not build wealth. He did not build stability. But he built presence. He built connections. He built a body of work that continues to speak long after he is gone. And that is the measure that matters most. Not what a man accumulates, but what he leaves behind, cannot be erased. Pete Carolan did not fit the mold. He did not need to. Because the mold was never designed to hold men like him.

THE MAN BEHIND THE PAINTED FROG

MS. ANN KENNEY , PHILANTHROPIST
PATRIOT SEAL HERITAGE FOUNDATION

I have always believed that certain people come into your life not just to be remembered, but to change the way you see the world. Pete Carolan was one of those people for me. Knowing him was not something casual or fleeting. It was something that stayed with you long after the conversation ended, long after the laughter faded. Pete had a presence that was impossible to ignore, yet it was never overwhelming. He was larger than life in all the right ways but grounded in a quiet humility that made you lean in rather than step back.

When I first met Pete, I did not fully understand the depth of who he was. I knew the stories, I had heard the "Pete facts," and I knew he was respected, but it was not until I sat across from him, listened to him speak, and watched the way he moved through a room that I realized he was something rare. He had this way of making you feel like you mattered, like your presence was important, even if you had just met him. There was no ego in him, no need to prove anything. He had already lived a life that most people could never imagine, and yet he carried it with such ease.

Over time, our friendship grew in a way that felt natural and steady. Every year, I would see him again, and it became something I looked forward to deeply. My daughter and my granddaughters would come with me, and Pete would spend time with them as if they were his own. He would sit with them, sketching on napkins, drawing his frogs, making them laugh with just lines of ink. Watching him with them was one of the clearest reflections of who he truly was. He had this ability to connect across generations, across experiences, across everything. Everyone who met him loved him, and it was not hard to see why.

Pete did not just tell stories—he lived them in a way that allowed others to feel them. When he spoke about his time in the Teams, about the men he served with, about the missions and the moments in between, there was always something deeper underneath it. He was not trying to impress anyone. He was sharing something real. And that authenticity stayed with you. It made you think differently about service, about sacrifice, and about what it really means to give your life to something greater than yourself.

He was a super patriot in every sense of the word. Pete loved this country in a way that was both fierce and deeply personal. It was not performative. It was not loud. It was woven into everything he did. You could see it in his artwork, in the way he spoke about history, in the reverence he had for those who served before him and alongside him. The Trident, the flag, the symbols of freedom were not just images of freedom to Pete, they were part of his identity.

The Trident, especially, carried a weight that he understood in a way few people could. It is more than an insignia. It stands for everything those men endure to earn it—the discipline, the sacrifice, the brotherhood, and the responsibility that comes with wearing it. Pete never saw it as something that ended when you took off the uniform. To him, it was something you carried for life. It was a promise to the men beside you and a commitment to those who would come after.

That understanding shaped the way he saw the world, and it shaped the way he treated people, especially veterans. Pete had deep compassion for those carrying invisible wounds. He understood that not all battles are fought in the open, and that the hardest ones begin after the mission is over. He did not need to ask questions to recognize when someone was struggling. He could feel it. And in his own quiet way, he would meet people where they were, offering connection, acknowledgment, and sometimes just the simple gift of being seen.

That is one of the reasons the mission of *The Patriot SEAL Heritage Foundation* speaks to me so deeply. When I learned more about what the Foundation was doing, it felt like something Pete would have stood behind without hesitation. Their mission—to help our heroes heal by providing K9 and equine therapy programs for Navy SEALs, veterans, and first responders facing PTSD, traumatic brain injuries, and other invisible wounds—is not plainly important, it is necessary.

I have seen enough, heard enough, and felt enough to understand that the transition from active service to civilian life is not something you can simply step into without consequence. These men are trained to operate at the highest level. They are conditioned to lead, to endure, and to survive under pressure that most people will never experience. But when that structure is gone, when the mission ends, they are often left navigating a completely different kind of reality—one that is quieter, more isolating, and often more difficult.

The wounds they carry are not always visible, but they are real. They affect relationships, identity, purpose, everything. And too often, those struggles go unspoken.

The Patriot SEAL Heritage Foundation is stepping into that space in a way that feels both powerful and deeply human. Their programs are not about quick fixes or surface-level solutions. They are about real healing—intentional, structured, and sustained.

The 3 Day equine therapy programs are something

that moved me deeply. Horses have a way of responding to what is real. They do not recognize rank, titles, or accomplishments. They respond to presence, to energy, to truth. Watching veterans interact with these animals, you can see something shift. There is a grounding that happens, a reconnection to self that does not require words. It allows them to rebuild trust—not just with the animal, but within themselves.

The K9 programs offer something equally powerful. The bond between a veteran and a dog is immediate and honest. It creates a sense of purpose, of responsibility, and of companionship that is incredibly healing. These dogs are not just companions, they are partners. They provide stability, comfort, and a sense of direction that can be life changing.

Supporting this work has never felt like a question for me. It felt like something I needed to do. I have donated thousands of dollars to *The Patriot SEAL Heritage Foundation* because I believe in their mission, and because I know the impact it has. This is not about charity in the traditional sense. It is about continuing the mission in a unique way. It is about standing behind the men who have stood for all of us.

When I think about Pete, I know this is something he would have cared about deeply. He never stopped honoring those who served. It was part of who he was. His art reflected it, his stories reflected it, and his presence reflected it. He carried that responsibility with him in everything he did.

I remember speaking to him days before he passed. There was something in his voice—a quiet knowing. There was no fear, just a sense of peace. When I got the message that he was gone, my heart broke in a way I cannot fully put into words. You do not meet people like Pete in a lifetime and this sudden loss felt like we had lost something much bigger than one person.

But in so many ways, I still feel him. I feel him in the places he loved, in the art he created, and in the people he touched. And I feel him in this mission.

The *Patriot SEAL Heritage Foundation* carries forward something that Pete lived every day. It carries forward the belief that brotherhood does not end when the mission is over. It carries forward the idea that we have a responsibility to take care of those who have given so much.

The need for this work is ongoing. It is growing. And it requires continued support from people who understand what is at stake. This is not something that can be sustained without involvement. It takes donors, sponsors, and individuals who are willing to step forward and be part of something meaningful.

The impact is real. It is not abstract or distant. It is seen in the lives that are changed, in the veterans who find their way back to themselves, and in the families who are restored along the way.

Pete Carolan's legacy, to me, is not just about what

he did , it is about how he made people feel. It is about the way he honored others, the way he loved this country, and the way he carried himself with quiet strength and humility. Supporting the Patriot SEAL Heritage Foundation is one way I can continue that legacy. It is one way I can ensure that the values he lived by continue to be effective.

I miss him. I miss his smile, his stories, and the way he could make everything feel meaningful. But I also know that what he stood for did not end with him. It lives on in the men he served with, in the art he created, and in the mission of organizations like this one.

The Trident is a symbol of strength, resilience, and brotherhood. But behind that symbol are real men—men who have carried extraordinary burdens in service to this country. They deserve more than recognition. They deserve support, seen, and they deserve to heal.

That is why this work matters. That is why I give. And that is why I will continue to stand behind this mission—for Pete, for the men he loved, and for the country he never stopped believing in.

THE PIRATE AND THE PATRIOT

STEVE MENSINGER, SIGMA CHI, FORMER MAYOR OF COSTA MESA, CALIFORNIA

The first time I met Pete Carolan at USC, he was not your typical college student. He had already survived the forge of BUD/S training and two tours in Vietnam. Pete arrived on campus as a seasoned Navy SEAL / UDT 13, a man who had already seen the world through the high-stakes lens of special warfare. He seemed to vibrate with a frequency the rest of us were still trying to tune into—a "loaded weapon" of a man who was strong, articulate, and possessed a raw energy that could either power a city or level a block.

He had arrived at Sigma Chi with a reputation and a nickname given to him by his BUD/S class 47 of 1968—those men who had been hammered into steel alongside him in the surf of Coronado. They called him

"Pete the Pirate." It was a title earned by capturing the pirate flag of his former college, Orange County Pirates. To the men in the Teams, Pete was a "hard charger" who embodied the ethos of the Trident: he was a man who would never give up, never retreat, and—most importantly—would never, ever "ring the bell to quit."

We became Sigma Chi brothers, bound by the White Cross. For Pete, this was not just a social club; it was a matter of lineage. The history of the "Duke," John Wayne, within Sigma Chi was what drew Pete to the fraternity. He saw in that bond a continuation of the "service over self" ethos he had practiced in the Navy. We had escapades throughout that first year that still leave me scratching my head and wondering to myself, *"did we really get away with that?"* After I graduated, we lost contact. I went into politics and eventually became the Mayor of Costa Mesa; Pete returned to a life of missions, art, and the occasional bit of "dust-kicking" havoc.

Our second chapter began in 2015, thirty years later. I was at a park for a ribbon-cutting ceremony when I looked over and saw a man painting a plane. There was something in the stance—a familiar, rugged intensity. "*Is that Pete Carolan?*" I asked. "*Oh yeah,*" the bystander replied. "*You know Pete?*" It had been three decades, but the bond of the Cross and the Trident remained as strong as ever. Just like that, the years evaporated between us, and we stepped back into step with me trying to place him into art projects with buildings I owned and ran.

His historic murals are still gracing those walls today.

What so many people did not realize was that Pete's artistic journey began in the most technical of environments. As a cartographer with UDT 13, Pete had a "spatial advantage" that most artists spend a lifetime trying to master. He understood terrain, depth, and the weight of a landscape. He viewed the world as a mission map, and his brain wired to solve complex problems without a formal plan, leaning instead into a raw, natural talent and a sudden spark of inspiration.

During the NASA capsule recovery days, Pete left his mark in a way only he could. He was the man who painted the infamous "Freddy the Frog"—the iconic UDT mascot—onto the side of the Sea King recovery helicopter used to retrieve the astronauts, an image so iconic it was recreated in the Apollo 13 movie. It was the perfect synthesis of his life: military precision meeting artistic expression. He was not just "painting a frog"; he was marking territory, building morale, and creating a legacy of service that reached from the depths of the ocean to the edge of space. I encourage everyone to visit his murals and support the foundation that preserves his work, ensuring his patriotism continues to inspire future generations.

Hiring Pete to paint for my city projects was not just a professional decision; it was a stewardship for his talent. Pete was a "mission guy" a trait forged in the Teams. There was a point to start and a point to finish, but the space in between was a battlefield of the mind, especially for him. He was the kind of American who thrived on the "impossible" task. If he did not have the right tools, he improvised; if he did not have a sketch, he let the history of the location guide his hand. He never backed down from a blank wall, viewing every surface as a chance to defend the American story.

Because of his background as a cartographer, he excelled in spatial scenes, and his legacy continues to inspire patriotism. His masterpiece, *"The Alamo,"* is still a testament to this gift. He did not just paint a building; he painted the gravity of the stand taken there. His work eventually reached the highest corridors of power.

Today, Pete's brushstrokes are preserved at the Pentagon, the White House, and the entrance of the National Navy UDT-SEAL Museum in Fort Pierce, FL, and the San Diego National SEAL Museum. The USS *Midway Museum* in San Diego, California. Across the country, his murals are on city buildings, firehouses, parks, VFWs, and military buildings, showing his deep patriotism for those currently serving and veterans of combat and victory for the United States.

His murals are loved inside the Coronado, CA, and epic SEAL TEAM waterholes, such as Blackie's and Mc.P's, which serve as anchors for the community, depicting fierce, unapologetic patriotism. His combat mural art has become a permanent part of the American landscape, ensuring that the legacy of those who serve is never forgotten. For these contributions, Pete was recognized with the Mayor's Award, a tribute to his status as a national treasure who never stopped fighting for his country.

To know Pete was to understand the war between the right and left sides of his brain. The SEAL in him demanded structure; the artist in him needed to bleed outside the lines. That friction often manifested in a struggle with alcoholism and mental health. He was a man of intense shadows, but he was also a man of immense light. He lived by the code that you address the problem head-on; you do not hide from it. *"Pete, keep trying,"* I would tell him. *"You're a better person when you address the person in the mirror."*

Art became his sanctuary, his way of healing the wounds that even a SEAL could not outrun. He eventually accepted Christ and would recount scripture with the same intensity he once used to recount tactical maneuvers. In the "fourth quarter" of his life, his art became his expression of faith and his ultimate contribution to the country he loved. He wanted to be remembered as a man who faced his demons with the same courage he faced an enemy, never ringing the bell on his own soul. He found peace in the work, creating beauty out of the chaos of his own experience, leaving behind a story of redemption that is as uniquely American as the flag he spent his life defending.

One of my most vivid memories is meeting him at McP's in Coronado—the legendary SEAL hangout. Pete knew everyone, and in that environment, he was the king of the "hard chargers." We were there for the dedication of a ship named for Michael Monsoor, a Medal of Honor recipient. Pete was in his element, surrounded by men who lived by the same "never give up" creed. He threw down words which led to a throw down of fists. The lines at times blurred for him as to which fight was the good fight. He was a man who could navigate a verbal war with international special forces one minute and present a deeply moving, hand-painted tribute to a grieving mother the next.

That was the Pete Carolan ethos: He got into places no one else could because he connected dots that others did not see. He gave himself no credit, always pointing the light toward others. When he gave a painting of horses to George W. Bush—one of his favorite pieces—it was not about the artist; it was about the recipient. He was a man who created solutions out of thin air, a patriot who used his gift to honor the sacrifices of his brothers-in-arms. He lived for the moment of the "gift," the moment when his art could serve as a bridge between the trauma of war and the peace of remembrance.

He gifted art to President Ronald Reagan while he was still the governor of California, to George W. Bush who kept his art and now hangs at his private residence. Donald J. Trump received art from Pete with a follow up letter thanking him for his service as a super patriot of the United States of America.

Pete Carolan's art legacy is a national treasure. He was a man who lived the Sigma Chi creed and the SEAL ethos simultaneously. He was uniquely American—a man who did what he said and said what he did. He was angry about 9/11, not out of hate, but out of a fierce, protective love for a country that welcomes all who wish to contribute. If he could have put the Trident back on and gone back to the front lines at seventy, he would have. Instead, he used his brush to defend our memory and created 911 murals and artwork showing our country's loss and grief, losing civilians and firefighters as well as first responders.

History will remember Pete for his authenticity. He was our Teddy Roosevelt—not always clean, often turbulent, but always moving toward a vision. He taught us that even if you are "wrong" sometimes, there is a valuable lesson in the struggle. Pete the Pirate has found his final harbor, but the murals he left behind and the oceans he swam and the heroes he admired—will remind us forever of what it means to serve. His life's work continues through the foundation he founded, the Patriot SEAL Heritage Foundation, ensuring that the "combat mural art" he created continues to inspire future generations of patriots as his works now tour our national museums and libraries as a time capsule.

To understand the spirit of "Pete the Pirate," you must stand before the scale of his vision. Pete did not just paint; he staged an assault on the canvas, capturing the spatial reality of the warrior's world. I encourage you to visit his works and walk the halls where his legacy lives on.

- National Navy UDT-SEAL Museum (Fort Pierce, FL): View his historic contributions to the UDT and SEAL heritage.
- The SEAL Museum (San Diego/Coronado, CA): Experience the art born in the heart of the Teams' home.
- Support the Mission: Join us in ensuring that the stories of our nation's hardest chargers are told through the inspired eyes of the Pirate himself.

Steve Mensinger, Former Mayor of Costa Mesa, CA 2015

Artist: Peter Carolan – 80th Grand Chapter Dedication to Sigma Chi Fraternity, USC.

LOS CABALLEROS
CATALINA ISLAND TREK
EST. 1978

THE HORSES REMEMBER

TOM CRONE, LOS CABALLEROS CATALINA COWBOY HERITAGE FOUNDATION

There are places in this world that change a man quietly. Not through spectacle or noise, but through silence, rhythm, and memory. Santa Catalina Island is one of those places. I first came to Catalina as a boy, back when the island's interior was still spoken about almost in whispers. Everyone understood there were boundaries. Tourists stayed close to the main town of Avalon while the deeper parts of the island remained protected, rugged, and mysterious. The interior belonged to the wild. Rolling hills stretched endlessly toward the Pacific. Dust trails wound through ancient canyons. Bison moved across the landscape like ghosts from another century. Even as a child, I sensed there was something sacred about that land. What I did not yet understand was that Catalina was not simply beautiful, it was healing.

Now, after more than twenty-five years riding Catalina Island with Los Caballeros, I can say without hesitation that the annual five-day ride remains the best week of my life every single year. Not because it is luxurious, but because it strips life down to its essentials. Horses. Brotherhood. Nature. Simplicity. Presence. In a world where people live disconnected from themselves, Catalina reconnects you. The moment we begin loading horses onto barges on the mainland, something inside me starts to shift. The farther we move from the coast, the quieter my mind becomes. Stress loosens its grip and the noise of life and stress disappears. Then the horses step onto the island, and the transformation begins.

People who have never bonded with a horse often misunderstand what these animals truly are. Horses are not simply transportation or recreation. They are deeply intuitive creatures that feel every emotion carried by the person beside them. They know when you are anxious. They know when you are grieving. They know when your spirit is unsettled. You cannot fake calm around a horse because the animal senses the truth of you immediately. That is why horses heal people. Long before anyone used the term "equine therapy," cowboys already understood the connection. Horses regulate the nervous system. They force you into the present moment. They teach trust, patience, and honesty. On Catalina Island, surrounded by nature and silence, that healing becomes even more powerful.

I have watched hardened men soften around horses. I have seen veterans, titans of industry, and those carrying years of invisible trauma begin breathing differently after only a few days on the island. Something changes when you wake before dawn, saddle your horse in the cold morning air, and spend an entire day moving through untouched country beside people who understand the value of silence. Conversations become more real out there. Men begin talking honestly around campfires beneath the stars. The walls people build around themselves on the mainland begin to fall away. Catalina has a way of reminding people who they were before life hardened them.

That is part of what made Pete 'the Pirate' Carolan fall in love with the island. Pete came into Los Caballeros through Jim Cunningham while I was serving at the board level. Becoming part of Los Caballeros is never casual. Membership involves a three-year vetting process because this is not simply a riding club. It is stewardship. Brotherhood. Responsibility. Respect for the land and the traditions that came before us and something you must be invited into. From the beginning, Pete understood that. Many of us already knew of Pete before he ever rode with us. A lot of the guys had spent time around the SEAL headquarters in Coronado and were familiar with his work honoring the Navy SEAL legacy. There was already an awareness surrounding him because excellence leaves a reputation long before introductions happen.

But what impressed me most about Pete was not his reputation. It was his devotion. Pete loved Catalina Island deeply and without condition. Every year he became one of the strongest supporters of the five-day ride. He poured his energy into preserving the culture surrounding it. Pete created one-of-a-kind artwork for the cowboys, pieces that captured the spirit of the island and the people who rode it. Those custom works were auctioned during events in Catalina to support the Conservancy, Los Caballeros, and the Catalina Cowboy Heritage Foundation. Pete understood something important: preservation requires participation. If you genuinely love something, you help protect it.

Behind Pete through all of it was Peggyann Morley, now the President of *The Patriot SEAL Heritage Foundation.* Everyone knew Peggyann was Pete's guardian angel. She supported him beyond belief, quietly and consistently, often behind the scenes where few people noticed. She kept him organized, grounded, and able to focus on the work and causes he cared most about. Every great man has someone standing beside him helping hold life together, and Peggyann was that person for Pete. In cowboy culture, loyalty matters deeply, and her loyalty to Pete never went unnoticed.

The spirit of Catalina's horseback culture lives

through people who understand the island's deeper purpose. You can see that in the owner of Catalina Island Horseback Adventures, Jeff Skelton and Blanny Avalon Hagenah, granddaughter of the Wrigley family involved in Catalina conservation and equestrian circles and Middle Ranch which is far more than a boarding facility. It is part of the island's living heritage. The ranch works closely with the Saddle Club and continues preserving horseback culture in a world increasingly disconnected from nature. Jeff accomplished something remarkable when he brought public horseback riding back to Avalon after nearly eighteen years without it. Eighteen years without horses available for the public to ride changed the energy of the island. You could feel the absence. Horses have been woven into Catalina's identity for more than one hundred years, and when they disappeared, something essential disappeared with them.

Bringing horses back restored part of Catalina's soul. Horses belong on that island. They always have. The history stretches all the way back to William Wrigley Jr. and Philip K. Wrigley, whose vision for Catalina was never about separating people from nature but connecting them more deeply to it. P.K. Wrigley believed in uniting the island through conservation, recreation, stewardship, and shared experience. That vision stays alive today through the Catalina Island Conservancy, which was created by the Wrigley family in 1972 and now manages nearly eighty-eight percent of the island.

David Solomon, the interim CEO of the Catalina Island Conservancy, understands something critically important about Catalina's future. One of the Conservancy's three core pillars is recreational use of the island. The horses are not separate from Catalina's history; they are central to it. Families like the Wrigley's and the Rusack's understand that preserving horseback culture means preserving part of the island's identity itself. The Rusack family oversees many of the businesses, tours, hotels, and lands connected to island life, including the property where Jeff Skelton's ranch operates. Everything on Catalina remains interconnected. The Conservancy, Middle Ranch, the Saddle Club, Los Caballeros, and the Catalina Cowboy Heritage Foundation all work together in different ways to protect the island's future.

Los Caballeros holds a unique place in that history because it remains the only group allowed to ride Catalina's interior in this way. That privilege was earned through generations of respect for the ecosystem, the flora and fauna, and the understanding that horseback travel allows people to experience the island without damaging it. Even transporting the horses is still a major operation. Every year, anywhere from sixty to 120 horses are brought over by barge from the mainland. Watching the horses unload onto Catalina never loses its magic. It feels timeless, as though history itself is stepping ashore.

For Pete, the five-day ride became deeply personal. The ride was never about spectacle for him. It was about brotherhood, silence, nature, and connection. Out there, removed from the distractions of modern life, Pete found peace among the horses and the people who understood him. Cowboy culture has always offered men something modern society struggles to provide belonging. Not performance or an image, real belonging.

When Pete passed away in August of 2022, the loss rippled through the entire Catalina community. Los Caballeros, Las Caballeras, Middle Ranch, the Saddle Club, Conservancy members, veterans, artists, and close friends all came together to honor him because they understood who he truly was. During the memorial, a Navy SEAL Trident was pounded into a wooden board as a permanent symbol of Pete's service and spirit. It was one of the most powerful moments I have ever seen because it was raw, honest, and deeply meaningful.

After Pete's death, the ride continued with what became known as the horseless rider. John Bennett, Pete's gunsel and a board member of the Patriot SEAL Heritage Foundation, arrived carrying enormous grief. But something extraordinary happened during that ride. The entire Los Caballeros community wrapped around John the way a herd protects one of its own. Nobody needed speeches. Nobody needed explanations. Through presence alone, the group made it clear that John was no longer standing outside the circle. He was now part of the herd forever.

That kind of brotherhood is becoming rare in modern life, but it still exists on Catalina Island. I often think about the connection between Navy SEALs, horses, cowboys, and healing. On the surface they seem like completely different worlds, but underneath they share the same foundations: discipline, trust, service, humility before nature, and the understanding that healing happens through the nervous system as much as through the mind. Today many retreats for veterans and operators are rediscovering what cowboys have known for generations. Horses, nature, and brotherhood heals people. Catalina creates the perfect environment for all three.

Every year when the ride ends and I leave the island behind; I carry part of it home with me. I remember the smell of horses at dawn, the sound of laughter around campfires, and the sight of riders moving across ridgelines against the Pacific sky. I remember watching strong men soften around sacred animals. I remember brotherhood. And every year, without fail, I begin counting the days until I can return again.

Because Santa Catalina Island is not simply a destination. It is a reminder that nature still heals people, that horses still carry wisdom, and that some traditions are worth protecting long after we are gone. The horses remember, and somehow, riding along with them across Catalina Island, we remember too.

PETE BECOMING A COWBOY

JIM CUNNINGHAM, MEMBER AND PAST PRESIDENT OF LOS CABALLEROS

I first met Pete Carolan in 2013 through my long-time friend, Mark Lackey. Mark called, and he wanted me to meet Pete. Now that got my curiosity because Mark had never talked about Pete before. Mark and I first met in the early 1980's. We enjoyed many adventures, shared our triumphs and our disappointments, entertained each other with spellbinding stories, and became remarkably close, trusted friends. Mark and Pete met when they were young men. Mark said that he had told Pete about my long participation in Los Caballeros and how much I enjoyed it. Pete asked Mark to make the introduction as he was interested in learning more about Los Caballeros.

Mark is truly a charming and multifaceted character. So, if Pete were a long-time friend of his, I would have wanted to meet him. I will never forget that first meeting with the three of us. I was in the company of two colorful personalities who were gifted storytellers. The banter between us was so fun. Pete intrigued me with his jovial personality, quick laugh, and boyish nature. He was dressed in full cowboy attire.

After explaining the Los Caballeros ride on Catalina Island and how it has enhanced my life, Pete said that he wanted to sign up. Then I told him the realities, it is a private men's western riding group. Anyone who wants to go on the Catalina ride must be sponsored by a member, and then the Board of Directors reviews their application. Even if he is accepted as a first-year guest, he is a guest for 3 years and only becomes a member if invited back. It is 5 days and takes a certain type of man. A guest must prove he belongs as a member. I told him about the steep cost because of the planning involved, barging everything to Catalina Island, the cost to have a horse rental provided, and the annual dues.

His facial expression changed to one of bewilderment, and his smile turned to a frown. The subject changed, and we continued to enjoy our beverages. Pete became quiet. I could tell that he was contemplating what I had said.

Suddenly, he looked directly in my eyes with determination and a smile, saying that he would do whatever it takes to be invited. That is what I wanted to hear. Then I asked about his riding ability and experience with horses, as that is an important qualifier. I could tell that stunned him. He sheepishly said that he had never been on a horse in his life but always wanted to learn. He shared that he was a cowboy at heart and saw this as his best chance to become one.

The belief of the American cowboy by the public has been shaped by movies, TV shows, books, and limited exposure to actual cowboys. The American cowboy is an American icon emulating true masculinity, toughness, camaraderie, being out in nature's elements, and loyalty. Men will dress the part to be identified with those qualities. That was Pete. As a Navy SEAL, I could understand why he wanted to become a cowboy, as there are similarities.

My first Catalina ride was in 1989. I was a young man and excited to be invited. I had never owned a horse but had riding experience. My dad put me on a horse as a young child, and I rode as often as I could. My uncles were great horsemen who owned several horses and taught me a lot about riding.

I came back from that first Catalina ride and bought my first horse. That led to several great friends and the cowboy lifestyle, doing cattle drives, brandings, competitions, horse training, and owning more horses. I became incredibly involved in Los Caballeros, becoming a Board member and then the President.

I knew if I decided to be Pete's sponsor, he would get accepted as a first-year guest, but he had to prove to me his serious intent. He did this by hounding me. He did everything to convince me to sponsor him with his vivid, interesting stories about his life and his inviting personality. He laid it on thick. Amongst his stories he told me about was his impressive competitive outrigger paddling history. At that time, I was a competitive outrigger paddler, so I understood the commitment to physical toughness. I could relate to the competitive spirit and the necessary teamwork for success.

I requested him to take me to a nice dinner with other Los Caballeros members who would support his application, and he had to pay for it. We did have that dinner, and in true Pete style, he somehow did not pay for it. That was the beginning of learning about Pete's talent for getting out of paying for things. Anyone who knows Pete well knows this about Pete. Somehow, he got away with it, and it did not seem to matter. It can be attributed to the entertainment value that he brought to every conversation.

I did sponsor Pete as a first-year guest for the Los Caballeros Catalina ride in 2013, and he was accepted. He was super excited when I told him. I told him I would arrange a rental horse for him, as I know the owner of the rental horses very well, and due to Pete's total lack of experience, I wanted to be sure he got a seasoned and safe horse. But when I said his acceptance was conditional on paying the full-ride fee and proving he had paid for a rental horse, it was obvious that he did not take that into consideration. I knew by then that it would be a big financial challenge for him. He told me he would figure out a way, and he did.

Once I got confirmation that Pete had paid for the rental horse, I called my friend who provides the horses. He has been on the Catalina ride several times and knows what it is like. It is not a leisurely ride. The trails are incredibly challenging. There are long hours in the saddle. There are riding competitions that all first-year guests must do, regardless of experience. I told him that the perfect horse for Pete was like a walking piece of furniture that he could not fall out of. I was a little worried about Pete, so I asked him to assign one of his wranglers to stay with Pete on the trails and watch over him. There was no way Pete could keep up with me as a seasoned rider.

There is a first-year guest orientation meeting to prepare them for what to expect and what they must do. Pete called me after with great enthusiasm and excitement. I heard that he became extremely popular with his first-year comrades from that meeting. The day came for all of us to board the chartered boat to Catalina Island. The departure time was 8:30 AM, and it was an hour away. I told Pete that he had to pick me up at 6:00 AM and drive us to the dock. *"No problem,"* he said. He was late and arrived in an old T-Bird coupe car full of stuff. There was no way we could fit in with what I needed to take and make it on time. I was pissed and chastised him. He did seem to care. We ended up taking my SUV. Looking back, he planned it that way.

From the moment we arrived at the dock, Pete acted like a kid who got to go to Disneyland for the first time. He smiled and laughed the entire time. His enthusiasm was infectious. His charming personality endeared him to the other riders. He was a horrible rider. Good thing, his horse was very forgiving, and Kyle, the assigned wrangler, was with him when a bull buffalo burst out of the brush and spooked the horses.

Pete, with no experience in how to sit in the saddle to be prepared for such events, was tossed and landed hard on his ass, per Kyle's retelling of the story. It was only the two of them. Kyle was amazed that Pete was able to get back on and continue to ride. Pete was proving to be tough. It did not change his gleeful demeanor. Pete fell off again the next day. This time in the arena, during one of the competitive events. It was a sloppy, involuntary dismount, confirming his lack of horsemanship skills.

He lay on the ground, not moving for a long minute. His gentle horse stopped, turned around, and looked down on him as if asking if he was okay. Horses are amazing, noble creatures that can sense the well-being and mood of humans. The horse turned and walked away, which I took as a good sign. Nevertheless, I was alarmed. The other men gasped with concern for Pete. I ran out to check on him. Lying motionless in the dirt with his eyes closed, he moved a little when he heard me coming. Suddenly, he looked up, smiled, and said, *"How did I do? Help me get up."*

He was fine but a little rattled. He walked out of the arena a little gimpy and had a cocktail followed by beers while sitting in the shade, not bothered by the fall. When it was time to ride back to camp, he insisted on riding back to camp and not going in a truck. I helped him mount his trusted, kind, and gentle steed, and off we rode together for 2 hours back to camp.

He acted like nothing was wrong. There was certainly nothing wrong with his ability to drink beers all the way back. He was up late that night having fun. He was determined not to miss a minute of the organic, spontaneous entertainment that happens late at night with the happily imbibed riders.

We returned to the mainland after the Catalina ride. We loaded my SUV and left the parking lot to find an attendant collecting the parking fee. Of course, Pete claimed that he had no money on him, so I had to pay. I insisted that he pay me back, and he promised that he would. When it became clear that he thought he could get away without paying me, I called our mutual friend, Mark Lackey. I told Mark that if Pete did not pay me back, he would not get invited back. Mark told Pete, and Pete panicked. It was just a few days after we returned from the ride when Pete called me wanting to come over to pay me back and bring steaks to BBQ. He obviously wanted to remain on my good side. He invited Mark to join us. We were having a fun time. Suddenly, he stood up and asked, *"Do you want to see my ass?"* Mark and I yelled, *"No!"* Pete smiled broadly, turned around, pulled his pants down, and bent over.

His ass was completely covered in blue, black, and green bruises from the spills he took off the horse. I could not fathom how he could ride the whole time with the pain he must have had. With all his experiences from being in Vietnam and as a UDT 13 Navy SEAL, it is easy to assume he has experienced more, and this was a minor inconvenience for him.

Pete was a consistent figure at all Los Caballeros events. The guys were always glad to see him and wanted him around to take part. I have other stories to tell about Pete, but they would be best told by others or Pete in his own words. So, I am not even going to try. Pete was a man who was difficult to describe in words. I am glad we became good friends. He had gifts that served him well, his engaging and charming personality that made him good friends quickly and lasted for years. His unique talent as an artist. His ability to tell entertaining true stories told with expression and joy. He could be difficult to manage at times, and he was quirky.

He approached life differently than anyone I have ever known, the good and the not-so-good. We did not agree completely on things, but he had a knack for growing on you positively. I think he did suffer from PTSD that affected his ability to properly navigate through life. But he seemed to figure out how to make the best of situations at times that he faced for his benefit. Being around horses and the cowboy way really was good for him. It is good for all of us. Ronald Reagan was known to say, *"There is nothing better for the inside of a man than the outside of a horse."*

I knew his health was failing from what he shared with me. I was aware he was in the hospital in 2022, and his immediate future was uncertain. It was a sad day when I learned that he had died. I went to the unofficial celebration of life for Pete at Blackies Bar in Newport Beach, CA. He was a constant fixture there for decades. The other regulars knew him well. Several of the regulars showed up to honor him.

Blackies, filled with people. We all went outside and stood in silence as the sun was dipping under the horizon, turning the clouds into different hues of warm colors. A man dressed in the formal Scottish Prince Charley Kilt suit played the bagpipes as the sun disappeared. Then everyone went back into the bar for more beers and multiple toasts in Pete's honor. It was heartfelt and quirky. The level of intoxication was obvious. Pete would have loved it.

There were two formal celebrations of life for Pete that followed. The first one was at Elks Lodge in Santa Ana, California, for those who could not attend the second one at the Naval cemetery in San Diego, where

Pete's remains are now. Followed by a reception on Coronado Island, where the SEAL Teams are headquartered. Both well attended. Asked to speak at each event, I was honored and had to accept.

When I stood up in front of the crowded room gathered on Coronado Island, I only knew few in the room. I remember saying that I do not know many of you here, but I know we all have one thing in common for sure. I told them all, *"We all bought Pete drinks."* Everyone laughed.

Pete became an American cowboy in spirit and by actions. I am proud that I had something to do with that. He told me that it changed his life by being around horses and the cowboy fellowship he so enjoyed. He never got the horsemanship figured out, but he never gave up. There was no quit in him. He is missed by me and many others in the Los Caballeros' brotherhood.

BENEATH THE CATALINA STARS

JEFF SKELTON, CEO CATALINA ISLAND HORSEBACK ADVENTURES

Horses have been an integral part of Catalina Island for more than 100 years and have been used to support transportation, working ranches, cattle operations, guided trail rides, and mounted staff. The island developed a genuine cowboy culture that many visitors today don't realize once existed at such a large scale. Philip Wrigley was once famously quoted as sharing that, *"the best way to see Catalina Island is on the back of a horse."*

Catalina Island sits just twenty-two miles off the coast of Southern California, yet the distance it creates is psychological more than geographical. Once you leave the mainland behind, something subtle but undeniable happens. The mental clutter that normally fills space begins to thin. The need to check phones, answer messages, stay ahead of obligations and it starts to lose authority. Out there, the world stops demanding attention every few seconds. Instead, it draws you in. Sitting around a campfire on the island, far from city lights, crowded streets, noise, and constant distraction, you are given something increasingly rare: uninterrupted presence with other human beings. And that changes everything because people stop talking about what they do, and they start talking about who they are.

In 2024, after nearly two decades without public horseback riding on Catalina Island, I had the vision and privilege to bring it back. When I launched Catalina Island Horseback Adventures, my goal was never to simply offer trail rides. I was interested in more than tourism in the traditional sense, or in packaging scenery into something consumable and forgettable. What I wanted was an experience—something slower, deeper, and more human than what most people are accustomed to in daily life. I wanted people to experience the Catalina I had come to know and love during my annual rides on the island with the Los Caballeros private riding group. A place where the noise of the modern world fades not gradually, but all at once, like a radio being turned down until only silence remains. A place where the horizon feels closer, the sky feels larger, and time itself begins to behave differently.

There is something about Catalina Island that resists urgency. Once you step onto it, the pace of the mainland feels almost artificial, like a language you forgot you were speaking. My intention was simple: bring people into that space, put them on horseback, and let the island do what it has always done best—strip life down to its essentials. Ride on the rugged trails. Move along vistas where the Pacific opens endlessly in the distance. Sit quietly in places where the only sounds are hooves, wind, and the distant cry of seabirds. Let people rediscover what happens when motion slows enough for awareness to catch up. What I didn't fully appreciate when I began riding horses on the island was how many extraordinary people those trails would bring into my life. Among them was Pete Carolan.

Sitting around a campfire overlooking the ocean on the backside of Catalina Island, I was reminded that some people in this world are simply built differently. Not in a way that announces itself loudly, but in a way that becomes obvious only after time has passed and observation has done its work. They have an aura shaped not by performance, but by a lifetime of lived experience—by hardship, discipline, humor, loss, and perspective earned the slow way, Pete was one of those men. A Los Caballero who found brotherhood in his fellow riders, much like he did in the TEAMS.

One of the first things I noticed about Pete was that he moved through the world differently than most people. During early trips with the Los Caballeros riders on the island, I watched the rhythm of camp life settle in as evening arrived. Horses were unsaddled and the gear was laid out. Conversations drifted into the firelight and live music from someone's acoustic guitar would rise and fall, eventually giving way to quiet. Most men, after a long day in the saddle, eventually returned to their tents, cots, sleeping bags, or air mattresses after a full day of riding and celebrating together. They followed the natural pull of rest, retreating into the comfort of routine even in the wilderness. What was clear was that Pete did not follow that pattern; he had his own rhythm of life.

Instead, I watched him pick up a blanket, drift away from camp without announcement, and disappear into the darkness beyond the firelight. There was no performance in it. No statement being made by him and it didn't feel like withdrawal or solitude for effect, it just felt like a personal preference. If you could find him you would see him curled beneath a tree, the sky open above him, ocean wind moving through the grass. He wasn't trying to endure anything. He wasn't "roughing it" in the way people sometimes frame outdoor experiences for storytelling purposes. He looked completely at ease. More than that—he looked to be exactly where he belonged. Looking back, that image stayed with me because it revealed something essential about him. Pete was not a man negotiating constantly with his environment. He did not need conditions to be ideal to feel comfortable. He simply adapted, accepted, and existed without unnecessary friction.

Over the next few days, I began to understand what made both Pete and the island so special. Life as a cowboy on Catalina has a way of dissolving the usual barriers between people. Conversations that might take weeks or months to develop elsewhere happen in a matter of hours on horseback. We spend long stretches riding through the island's interior—through canyons, over ridgelines, across stretches of terrain that feel untouched by time. The horses set the rhythm. The land dictates attention. Laughter and cowboy camaraderie fill the gaps of silence. And because of that, conversation becomes something more natural, less constructed. Around the campfire at night, those conversations deepen further. What makes them different is not just the setting, but the absence of ego-led performance by anyone there, no competition for attention, just comradery. Professional accolades and accomplishments rarely matter in that space. It is not that careers are irrelevant, but that they are no longer the center of gravity. Titles stop shaping perception of people and the external markers of identity fade into the background. What remains are the things that cannot be listed on paper.

We poke fun at each other, talk about horses, music, family, faith, friendship, loss, fear, disappointment, and the quiet hopes people rarely say out loud. We talk about the weight of life without turning it into something theatrical. And in that environment, something subtle happens and honesty becomes easier. Each person knows he can speak without being corrected, measured, or reduced. He knows he will be heard without interruption. And sometimes, that is enough to reset and recalibrate. Sometimes no advice is needed, just the presence of someone listening fully, without an agenda.

A few days into one of those early trips, Pete mentioned that he had served in the Navy as a UDT 13 Frogman in Vietnam. He said it the same way he said everything else—without emphasis, like it was simply part of the story, not the defining chapter of it. I knew enough to recognize that such service represented something extraordinary, but I also knew I didn't yet understand what it meant to live it. So, I listened as he spoke about his time in service with the same humility that seemed to define everything about him. There was no attempt to elevate himself, no desire to impress anyone, no sense that he was revisiting those years for validation. He was simply describing pieces of a life that had happened. At one point, I asked him—naively, in hindsight—about his favorite mission as a Navy SEAL / UDT 13. He paused briefly, then said something unexpected. Pete told me, *"Apollo 13 capsule recovery mission in 1970."*

That single answer shifted the entire tone of the conversation. I remember sitting there, fire cracking in the background, ocean wind moving through the trees, listening as Pete began to describe his role as dive team leader during the recovery of the Apollo 13 capsule. The world held its breath as the astronauts were hurling back to earth's gravity and Pete was right there in the middle of the ocean waiting for them.

Not one to rest idly while waiting, he shared the story of how he and his SEAL buddy Al Starr snuck down the bay of the Ticonderoga where the helicopter that was to be used for the capsule recovery was. He had brought air brush paint cans and a cardboard cutout of the now infamous "Freddy the Frog" with him. He told us, *"I wanted to make our mark. UDT 13 was here."* Without permission or fanfare, he and Al Starr ventured out on a covert escapade to the helicopter and painted "Freddy" boldly on the side of it. We all roared around the fire listening to him tell the story as if defacing military equipment were "normal" and picturing Al Starr holding up the cardboard cutout of Freddy, like contraband, would actually work out for him. He shared with a snicker and smile how leadership was originally less than thrilled at his handiwork.

There are moments in life where you become aware that you are hearing something directly from history itself—not filtered, not summarized, not interpreted—but spoken by someone who was there when it unfolded. This was one of those authentic moments that became a shared memory in a brotherhood. No documentary, no film, no written account could match the experience of hearing it in his voice, with his pacing, his pauses, and the quiet matter-of-fact way he described things that most people will only ever encounter through archival footage or textbooks. It was not dramatic in the way Hollywood would make it, it was real and that is what made it more powerful for us all.

That conversation marked the beginning of a friendship that would continue for years. Over time, I shared countless hours with Pete and many other remarkable men—on horseback, on beaches, and around campfires overlooking the Pacific. What developed among us was not simply friendship in the casual sense, but something closer to brotherhood. It was not declared; it was formed slowly and naturally. One of the most remarkable aspects of those relationships was how effortlessly they resumed after long periods of separation. Months could pass without contact. Sometimes a year. Yet when we returned to the island and saw each other again, there was no sense of distance to repair.

The same thing happened on the trail. Conversations would begin in one location, often in the middle of a ride, only to be interrupted by terrain, wildlife, or the simple logistics of moving horses safely through difficult ground. A bison might appear in the distance, pulling attention away. A steep section of trail might require focus and silence. Someone might announce a humorous "man down" as a rider stepped off briefly to pick up a lost hat, make a saddle adjustment or to take a quick bathroom break. And then the conversation would stop—mid-thought, mid-sentence, mid-story.

Two days later, on an entirely different trail, someone would ride up alongside you and say, without hesitation, *"So as I was saying…"* And just like that, it would continue. That rhythm taught me something about connection. That real relationships do not require constant maintenance in order to remain intact. They require presence when they exist—and trust when they are not actively being expressed.

Over time, I began to see Catalina not just as a place of adventure, but as a place of restoration. People arrive on the ferry carrying things they rarely articulate—stress, grief, anxiety, trauma, responsibility accumulated over years. And something shifts when they are placed in an environment where they are no longer managing information, schedules, or expectations. Horses are a major part of that shift. They do not respond to language in the way humans do. They respond to energy, tension, calm, and intention. They require honesty. There is no convincing them otherwise. If you are anxious, they know it. If you are grounded, they respond to it. They bring people into the present moment not by instruction, but by necessity. Combined with the natural environment of Catalina, that creates something profoundly stabilizing.

Those observations eventually helped shape a partnership between Catalina Island Horseback Adventures and the Patriot SEAL Heritage Foundation. Together, we began developing equine-based retreats for members of the Naval Special Warfare community. These experiences combine horseback riding, therapeutic support, guided reflection, and immersion in the natural world of Catalina Island.

The purpose of these retreats is not to prescribe healing, it is to create conditions where it can occur, a starting point. For Navy SEALS navigating the effects of combat, PTSD, traumatic brain injuries, and the complex transitions that follow service, that environment matters.

Horses slow everything down. Nature reduces noise. Brotherhood reintroduces the feeling of being understood without explanation. And in many ways, those same qualities were present in Pete. When I think of him now, I return to that image—Pete sleeping beneath the stars while the rest of camp settled into tents. It was not just a moment of rest; it reflected character. A man unburdened by the need for unnecessary structure. A man who could exist fully in whatever environment in which he was placed . The island suited him for that reason because it stripped everything down and in doing so, it revealed what mattered.

Over time, I have come to believe that is what Catalina continues to do for everyone who spends real time there. It removes enough of the modern world's noise that people begin to hear themselves more clearly. It creates space for conversations that would otherwise never happen. It fosters friendships that outlast seasons.

It reminds us that meaning is rarely found in complexity. It is found in simplicity, attention, and shared experience.

For me, Pete Carolan will always be part of that story. Not as a highlight or a footnote, but as a presence woven into the memory of the place itself. A reminder that the most meaningful encounters in life are not always planned, and the most lasting impressions are often made quietly, without intention. He was one of those rare individuals whose presence elevated every conversation, lightened every mood with his tales of victory in his warrior stories, everyday life observations, sea faring tales of his boat, *The Patriot SEAL* and deepened every experience with his ability to show up and remain present.

Through him, and through many others who have shared these trails and campfires, I have come to understand something that continues to shape everything we do on that island: The greatest adventures are not always defined by where you go. Sometimes they are shaped by who you are with, the season of life in which you arrive there, and the perspective you carry when you leave. Sometimes they are found on horseback, beneath an open sky and far from the noise of the world. And sometimes the great adventures are found in the company of a friend who reminds you - without ever needing to say it out loud - that life is simpler, and more meaningful, than we often allow ourselves to believe.

THE LAST STORY OF LEGACY

HUNTER RUSACK OF THE WRIGLEY FAMILY

Every great story eventually returns to its roots. Throughout *The Painted Frog,* readers have journeyed through the remarkable life of Pete Carolan—Navy Frogman, UDT warrior, artist, horseman, mentor, and friend. His story spans battlefields, oceans, and decades of service, but some of the most meaningful chapters were written far from war, on the trails and ridgelines of Catalina Island. As this book draws toward its conclusion, it is fitting that one of the final voices belongs to a man whose family legacy has been interwoven with Catalina for more than a century.

Hunter Rusack, a fifth-generation Catalina Island descendant of the Wrigley family, represents the continuation of the traditions that helped shape the island Pete loved so deeply. Through his family's stewardship of Catalina, their role in developing the island's Arabian horse program, and their connection to the enduring brotherhood of Los Caballeros, Hunter carries forward a legacy rooted in preservation, community, and shared experience.

What makes this contribution especially meaningful is that Hunter never personally knew Pete Carolan. Yet Pete's name continues to live within the culture of Catalina and Los Caballeros. As Hunter notes, Pete is still *"one of the most talked-about riders in Los Caballeros history,"* remembered not for seeking recognition, but for the humility, authenticity, and quiet leadership that defined his character. In many ways, that observation captures the essence of Pete's legacy. Legends are not created by titles alone. They are created by the stories others continue to tell long after a person has gone. They are sustained through respect earned, friendships forged, and the example left behind.

Hunter speaks throughout this interview about the values that have always defined Catalina Island: stewardship, brotherhood, leadership through calmness, connection to the land and sea, and the belief that meaningful relationships are built through shared experiences rather than words. These same principles shaped Pete Carolan's life.

The Patriot SEAL Heritage Foundation is honored to include Hunter's perspective in these pages because it serves as a bridge between generations. His reflections remind us that legacy is not something preserved in museums alone. It lives through traditions. Through communities. Through stories passed from one person to another. As the Foundation prepares to bring Navy SEAL veterans back to Catalina Island through equine and ocean-based healing retreats, Hunter's words reinforce a truth Pete understood well: healing, brotherhood, and purpose are often found in the simple act of gathering together—in the saddle, on the water, around a table, or along a trail.

Catalina Island has always been a place where landscape, history, horses, and community converge. It is still a place where connections are forged naturally and where stories continue to take root. For that reason, and for his willingness to contribute to preserving this important history, the Patriot SEAL Heritage Foundation extends its sincere gratitude to Hunter Rusack and the Rusack-Wrigley family. This final chapter is more than an interview; it is a testament to continuity. A reminder that Pete Carolan's story did not end with his passing. It lives on in the island he loved, in the traditions he helped strengthen, and in the people who continue to carry those traditions forward. The pages that follow belong to that continuing legacy.

With gratitude and respect,
The Patriot SEAL Heritage Foundation
"Preserving the Legacy. Inspiring the Future."

CATALINA ISLAND & LOS CABALLEROS

HUNTER RUSACK

Catalina Island has a long history shaped by its geography, its residents, and the traditions that have developed over generations. Among those traditions is Los Caballeros, a private and esteemed horseback riding group that has been part of island culture for more than seventy-five years. The organization remains connected to many of the families and individuals who have contributed to the island's history through ranching, conservation, tourism, and community involvement.

This conversation with Hunter Rusack reflects that perspective. A fifth-generation steward of the island's legacy, whose family's history remains deeply intertwined with its culture and traditions. Hunter grew up around many of the traditions that continue to define the island today, including horseback riding, outdoor recreation, and community events. His observations come from personal experience and family history rather than historical study alone.

Although Hunter never met Pete Carolan personally, Pete's name continues to come up in conversations among members of Los Caballeros and others connected to Catalina. He remains a recognized figure within that community because of his military service, artwork, and membership in the Los Caballeros.

The discussion that follows focuses on Catalina Island, the history of Los Caballeros, the role of horses in island culture, and the practical realities of preserving traditions while continuing to welcome new generations of visitors and residents.

Dr. (h.c.) Lisa Ray: Hunter, let's start with your family's history on Catalina. How far back does it go?

Hunter Rusack: *My great-great-grandfather, William Wrigley Jr. purchased Catalina Island with a group of investors in 1919. He developed much of the tourism infrastructure we see on the Island today. His son, my great-grandfather, Philip Knight Wrigley, carried on his legacy and went on to build the Airport in the Sky, El Rancho Escondido, and develop the Arabian horse program on the island. His love for both horses and Catalina led to his involvement in the early formation of Los Caballeros, which started around 1950—so we're talking about a 76-year tradition at this point.*

As a child, I'm not sure I truly appreciated the unique history and legacy my family left behind. For me, it was just Catalina. As I got older, and especially as I've taken on a full time role on Catalina, I started to appreciate and understand the history and its impact.

Dr. (h.c.) Lisa Ray: What are your earliest memories of riding horses on the island?

Hunter Rusack: *There are some photos of me from when I was about two years old. They were taken near El Rancho Escondido, for Arabian Horse World magazine. I don't remember any of it, but the photos are pretty fun to look at. Later, when I was six or seven, my grandfather gave me a pony named Stormy. Despite only seeing Stormy about once a month on our pilgrimages to the Island, that was my first real exposure to horsemanship.*

To be honest, I wasn't particularly interested in horses growing up. I was more interested in doing whatever kids do. It wasn't until I started participating in Los Caballeros rides, mostly to spend time with my brother, that I became more interested in horsemanship and riding. It's gone from something I did a couple times a year to a regular part of my life.

Dr. (h.c.) Lisa Ray: Tell me about Los Caballeros. What makes it significant to you?

Hunter Rusack: What makes it unique is the variety of people who participate. You have people from different professions, backgrounds, ages, and different parts of the country. Once everyone is on horseback for a few days, those distinctions matter a lot less.

We spend time together, ride, eat, drink, and have conversations that probably wouldn't occur in our normal day-to-day lives. That's part of what has kept it going for more than seven decades.

Dr. (h.c.) Lisa Ray: What's your most memorable experience with Los Caballeros?

Hunter Rusack: *One that stands out was a ridgeline ride with my brother Austin. It was our first year on the ride and we had set off on our own not knowing that we were jumping into the most technical ride of the trip. Along with the dramatic reviews comes steep terrain, and the ride required paying attention to every step the horse was taking. We worked our way through the trail together and ended up having a great time.*

We've done that trail since then, but that first effort stands out to me. I learned a lot about the capabilities of the horses in that moment and the importance of staying relaxed in the saddle.

Dr. (h.c.) Lisa Ray: What have you learned about leadership through horseback riding?

Hunter Rusack: Horses require consistency. If you're unpredictable or not paying attention, they'll let you know pretty quickly. They really do pick up on negative energy, fear, or a lack of confidence. If you're freaking out, they will probably respond. You have to stay calm, make deliberate decisions, and try to avoid being erratic.

You also learn that every horse is different. The outfitters do a good job matching riders with horses, but once you're out there, you have to adapt to the horse you're riding. There are some similarities to working with people. Not everyone responds the same way. You have to pay attention, communicate clearly, and make adjustments when necessary.

Dr. (h.c.) Lisa Ray: How does experiencing Catalina by horseback differ from other ways of seeing the island?

Hunter Rusack: *Horseback gives you access to parts of the island that most visitors never see. You're covering terrain differently than you would in a vehicle or on foot, and you spend more time paying attention to what's around you because you're responsible for the horse and the trail. It's one of the cooler ways to see what Catalina has to offer. For a brief moment you can forget that the craziness of the real world is only 26 miles away.*

Dr. (h.c.) Lisa Ray: Let's talk about the return of horses to Catalina more broadly. How has the community reacted?

Hunter Rusack: *Overall, the reaction has been great! When Jeff Skelton brought horseback riding back into public view through Catalina Island Horseback Adventures, there was a lot of interest. Horses have been a big part of Catalina's history, so for many people it felt like a natural fit. Like anything, there are different opinions, but most people recognize horses as an important part of the island's identity and Jeff has done a really good job engaging the community.*

Dr. (h.c.) Lisa Ray: You work at the Island Company. How do you think about balancing tourism with preserving the island's character?

Hunter Rusack: *It's always a fun challenge. The island depends on tourism; at the same time, there is a responsibility to maintain the balance that makes Catalina unique.*

The main goal is to preserve the history, legacy, and tradition of the Island and Island Company while keeping things fresh and inviting for those who come to visit.

Dr. (h.c.) Lisa Ray: What do you see for the future of Catalina over the next 20 years?

Hunter Rusack: *I think the overall approach will remain pretty similar. Improvements and updates to our operations if they make sense, but preservation of Catalina's charm will continue to be a priority. We are fortunate to get to do what we do and it's all about building an amazing experience for all those who come to visit.*

Dr. (h.c.) Lisa Ray: We're planning to display Pete Carolan's artwork at the 67th Annual Catalina Island Art Festival and at the Catalina Museum during the summer of 2027. How does that fit into honoring the island's legacy?

Hunter Rusack: *I think it makes sense. Even though I never met Pete personally, his name comes up frequently in conversations with members of Los Caballeros. He was clearly someone who left an impression on people. His art is great and displaying it gives visitors another way to learn about someone who had a big impact on Los Caballeros and all of us who are associated with it. Beyond that, it sounds like Pete was an amazing man and he should be recognized for everything he did for our country. His artwork is a cool way to do that.*

Dr. (h.c.) Lisa Ray: Let's talk about Pete Carolan himself. What was your impression of him from what you have learned?

Hunter Rusack: *From everything I've heard, he was a humble, respected military figure who made a real impact on the people around him. He seems like someone who didn't try to make things about himself, which is probably why people remember him so strongly. People talk about his military service, his involvement with Los Caballeros, and the relationships he built over the years. The fact that people continue talking about him today probably says as much as anything else.*

Dr. (h.c.) Lisa Ray: ***The Patriot SEAL Heritage Foundation*** **is developing a three-day equine retreat for Navy SEALs. The first of many is scheduled for September 21–23 right here on Catalina. What do you think is essential for making that program successful on Catalina?**

Hunter Rusack: *I think the most important thing is giving people opportunities to spend time together. Horseback riding is part of that, but so is the ocean, meals, conversations, and activities that get people out of their normal routines. In my opinion, there are few better ways to destress and be present than a weekend outing to Catalina.*

Dr. (h.c.) Lisa Ray: Why is the water experience during the retreat so important for a Navy SEAL?

Hunter Rusack: *The ocean is an important part of the Catalina Island experience for everyone. Whether someone chooses to dive, snorkel, fish, or simply spend time on the water, it's part of what makes the Island so special. The beaches all over California are amazing, but the water seems clearer and more abundant on Catalina.*

It's great for everyone, but I can imagine it would be especially appealing to veterans who have spent a lot of their lives around the water.

Dr. (h.c.) Lisa Ray: Before we wrap up—what's the story behind buffalo milk?

Hunter Rusack: *[Laughs] That's a Catalina classic. A bartender in Two Harbors came up with it back in the 1970s. It's become so popular that you will now see some variation of the cocktail at most bars on the Island. While delicious, it's essentially an adult milkshake, so it's hard to drink more than 1 or 2.*

Dr. (h.c.) Lisa Ray: Final question. What do you hope people understand about Catalina Island and its connection to this community?

Hunter Rusack: *Catalina has a long, rich history, and countless people have contributed to what it is today. Whether it's the Island Company, the Catalina Island Conservancy, Los Caballeros, local businesses, or the people who live here year-round, everyone has a role to play. I think people return because Catalina offers a departure from day to day life, and unique experiences that can't be found easily elsewhere.*

ACKNOWLEDGEMENTS

The Patriot SEAL Heritage Foundation is honored to present *The Painted Frog* as a tribute to the extraordinary life, service, and legacy of Pete Carolan. While this book celebrates one remarkable Frogman, it also stands as a testament to every member of the Underwater Demolition Teams and the Navy SEAL community whose courage, sacrifice, and quiet professionalism helped shape our nation's history.

We extend our sincere appreciation to the volunteers, supporters, donors, and strategic partners who believe that preserving these stories is both a responsibility and a privilege. Your continued commitment ensures that the history of our nation's elite warriors will not be forgotten and that future generations will understand the values of honor, humility, service, and sacrifice that define them.

We are especially grateful to our strategic partners, including Fast Passport Centers and Catalina Island

Horseback Adventures and nonprofit Veteran Canine Intelligence Academy whose support strengthens our ability to preserve military history while expanding programs that serve veterans and first responders through healing, education, and outreach.

To the museums, historians, researchers, archivists, photographers, and collectors who safeguard these invaluable pieces of our military heritage, thank you for helping preserve the artifacts, photographs, documents, and memories that make books like this possible.

Finally, we acknowledge every veteran, active-duty service member, first responder, military family, and patriot who continues to support our mission. Your dedication reminds us that preserving history is not simply about remembering the past—it is about inspiring the future.

May Pete Carolan's legacy continue to encourage courage in the face of adversity, selfless service to others, and an unwavering love of country.

PATRIOT SEAL HERITAGE FOUNDATION PRESIDENTIAL LETTERS

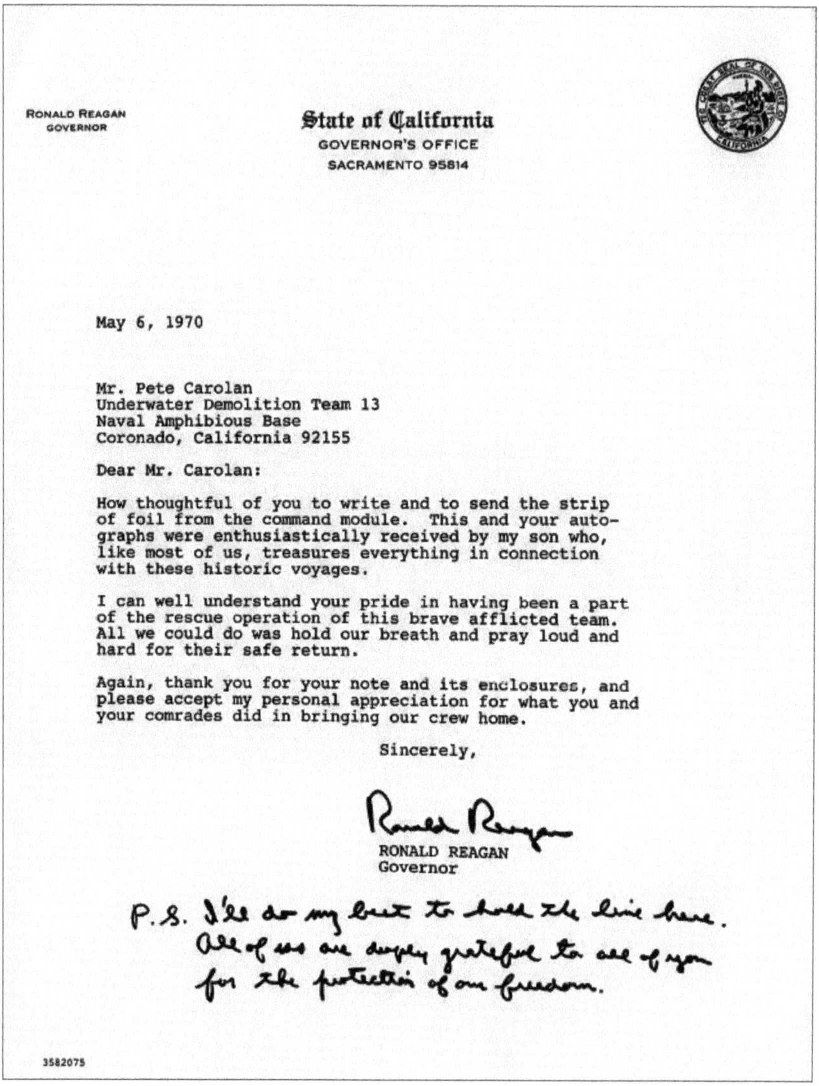

RONALD REAGAN
GOVERNOR

State of California
GOVERNOR'S OFFICE
SACRAMENTO 95814

May 6, 1970

Mr. Pete Carolan
Underwater Demolition Team 13
Naval Amphibious Base
Coronado, California 92155

Dear Mr. Carolan:

How thoughtful of you to write and to send the strip of foil from the command module. This and your autographs were enthusiastically received by my son who, like most of us, treasures everything in connection with these historic voyages.

I can well understand your pride in having been a part of the rescue operation of this brave afflicted team. All we could do was hold our breath and pray loud and hard for their safe return.

Again, thank you for your note and its enclosures, and please accept my personal appreciation for what you and your comrades did in bringing our crew home.

Sincerely,

Ronald Reagan

RONALD REAGAN
Governor

P.S. I'll do my best to hold the line here. All of us are deeply grateful to all of you for the protection of our freedom.

3582075

THE WHITE HOUSE

WASHINGTON

September 25, 2003

PO2 Pete Carolan, USNR
Post Office Box 181422
Coronado, California 92178

Dear Petty Officer Carolan:

Thank you for your letter and painting, "Let Freedom Ring, From Sea to Shining Sea." I appreciate your kind gesture and thoughtfulness.

America is inspired by the courage, dedication, and sacrifice of our men and women in uniform. Our Nation is grateful for your service.

Laura joins me in sending our best wishes. May God continue to bless America and all who defend her.

Sincerely,

George W. Bush

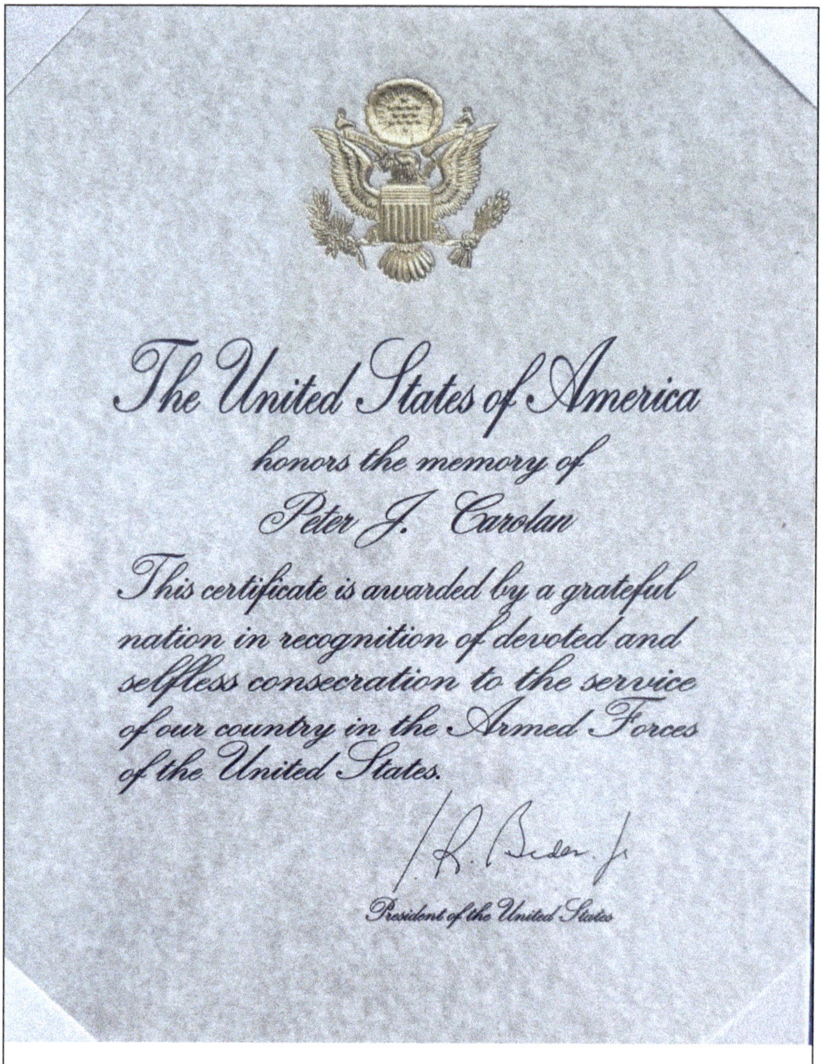
The United States of America
honors the memory of
Peter J. Carolan
This certificate is awarded by a grateful nation in recognition of devoted and selfless consecration to the service of our country in the Armed Forces of the United States.
President of the United States

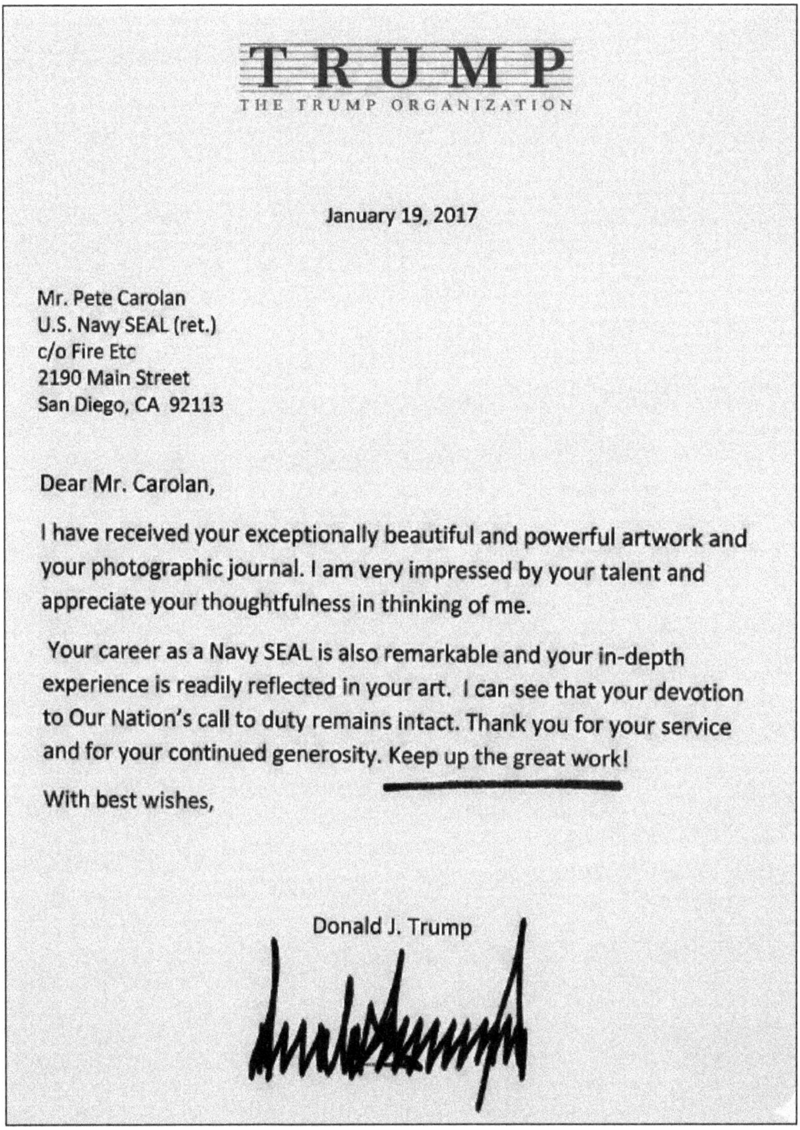

TRUMP

THE TRUMP ORGANIZATION

January 19, 2017

Mr. Pete Carolan
U.S. Navy SEAL (ret.)
c/o Fire Etc
2190 Main Street
San Diego, CA 92113

Dear Mr. Carolan,

I have received your exceptionally beautiful and powerful artwork and your photographic journal. I am very impressed by your talent and appreciate your thoughtfulness in thinking of me.

Your career as a Navy SEAL is also remarkable and your in-depth experience is readily reflected in your art. I can see that your devotion to Our Nation's call to duty remains intact. Thank you for your service and for your continued generosity. Keep up the great work!

With best wishes,

Donald J. Trump

ACCOMPLISHMENTS

Navy UDT-SEAL Museum
3300 North A1A
North Hutchinson Island
Fort Pierce, FL 34949
www.navysealmuseum.com
Tel: 772-595-5845

PRESS RELEASE

FOR IMMEDIATE RELEASE

Contact: Rolf Snyder
Office: 913-642-5134
Cell: 913-244-4782

SILENT AUCTION TO FEATURE FAMOUS ARTIST

Fort Fierce, FL October 22, 2006 – The Navy UDT-SEAL Museum's annual Silent Auction will feature works of art by Pete Carolan. Pete Carolan's art has been featured in films such as Ron Howard's "Apollo 13", and is also worn by the crew of the International Space Station. The Silent Auction, which is a significant fundraiser for the Navy UDT-SEAL Museum, will take place on November 9-12, 2006 during Muster XXI.

Similar artwork by Pete Carolan has been sold for as much as $17,000 at other similar auctions. His posters and paintings regularly fetch $1,000 a piece. A source close to the Museum said "Mr. Carolan's generous donations are the flagships of this year's Silent Auction. It is a real honor to have such a famous artist take notice of us."

Pete Carolan, a U.S. Navy SEAL and member of the UDT-13 Recovery Crew, was indispensable in the evolution of "Freddy the Frog", the mascot of the Underwater Demolition Teams. Pete, without prior authorization, painted Freddy the Frog on Recovery Helicopter #66, which would recover the Apollo 13 crew. He was saved from certain disciplinary action when the Recovery Helicopter's commander. CAPT. Smiley, remarked that he liked the frog on his bird. Recovery Helicopter #66, is currently on permanent display aboard the USS Midway, now a museum in San Diego, California.

The Navy UDT-SEAL Museum is the only museum dedicated to preserving the history of the Navy SEALs and their prodecessors, including the Underwater Demolition Teams, Naval Combat Demolition Units, Office of Strategic Services Maritime Units, and Amphibious Scouts and Raiders. Located in Fort Pierce, Florida – the birthplace of the Navy Frogman – the Museum promotes public education by providing the opportunity to explore the history of the Navy SEALs in an atmosphere of respect and honor.

###

If you would like more information about this topic please contact Rolf Snyder at 913-642-5134 or per email at Rolfs00@web.de.

FIRST FLEET

The Secretary of the Navy takes pleasure in presenting the Navy Achievement Medal to

Airman
Peter Joseph CAROLAN
United States Navy

for service as set forth in the following

CITATION:

"For professional achievement in the superior performance of his duties while serving as a member of the Underwater Demolition Team THIRTEEN detachment embarked in USS IWO JIMA (LPH-2) for the Apollo THIRTEEN recovery operation from 31 March 1970 to 25 April 1970. As a member of the swim team which executed the recovery of the astronauts and command module, Airman CAROLAN performed his assignments flawlessly. He executed his tasks in proximity to the command module with rare skill and precision, ensuring the proper installation of the flotation collar and egress raft for the debarkation of the astronauts. Airman CAROLAN's resourcefulness, professional skill and devotion to duty reflected great credit upon himself and were in keeping with the highest traditions of the United States Naval Service."

For the Secretary of the Navy

RAY PEET
Vice Admiral, U.S. Navy
Commander FIRST Fleet

UNITED STATES NAVY COMMANDER FIRST FLEET

CHIEF OF NAVAL OPERATIONS

The Secretary of the Navy takes pleasure in presenting MERITORIOUS UNIT COMMENDATION to

UNDERWATER DEMOLITION TEAM THIRTEEN

for service as set forth in the following

CITATION:

For meritorious service while participating in combat operations against enemy forces in the Republic of Vietnam from 1 March to 27 August 1969. As a member of the Amphibious Force, United States SEVENTH Fleet, Underwater Demolition Team THIRTEEN provided vital combat support against heavily defended enemy military and logistic installations and lines of communication. Under constant threat of enemy ambush, hazardous surf, and dangerous marine life, the Team conducted combat swimmer hydrographic reconnaissance of over seventy-five miles of enemy beaches along the coastal region of the Republic of Vietnam. The results of these operations provided vital hydrographic and backshore intelligence required to enable the amphibious force to land allied troops on hostile beaches in order to search out and destroy the enemy. During the same period, the members of Underwater Demolition Team THIRTEEN successfully participated in over one hundred combat riverine operations with SEA LORDS and SEA TIGER forces, in which they experienced numerous enemy-initiated ambushes. In spite of combat casualties, they carried the battle to the enemy, destroying substantial numbers of enemy fortified positions, structures, transport craft, and food caches. Underwater Demolition Team THIRTEEN enhanced river safety by the destruction of river-bank fortifications, opening previously-denied waterways to military patrol craft and civilian commerce. The outstanding courage, resourcefulness, and professional performance demonstrated by the officers and men of Underwater Demolition Team THIRTEEN were in keeping with the highest traditions of the United States Naval Service.

All personnel attached to and serving with Underwater Demolition Team THIRTEEN during the above-designated period, or any part thereof, are hereby authorized to wear the Meritorious Unit Commendation Ribbon.

For the Secretary,

T.H. Moorer

T. H. Moorer
Admiral, United States Navy
Chief of Naval Operations

DEFENDING KOREA'S FREEDOM

NAVAL SPECIAL WARFARE **FONT: STENCIL**
*** GROUP ONE ***

Presented To:

SO1 PETER J. CAROLAN
FOR PARTICIPATION IN OPERATION FOAL EAGLE
IN SUPPORT OF THE DEFENSE OF THE REPUBLIC OF KOREA,
MARCH THRU APRIL 1ST 2007

FONT: (ALGERIAN)

CAPT. Robert E. Rochfort, USN
NSWG Task Force Commander

Font: (Times New Roman)

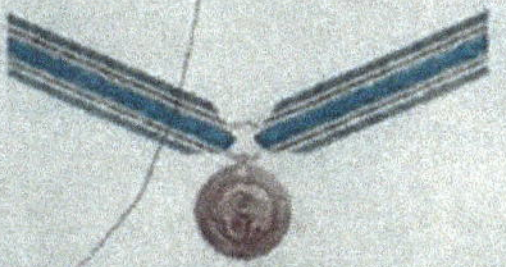

THE UNITED STATES OF AMERICA

TO ALL WHO SHALL SEE THESE PRESENTS, GREETING: THIS IS TO CERTIFY THAT
THE SECRETARY OF DEFENSE HAS AWARDED THE

KOREAN DEFENSE SERVICE MEDAL

TO

SERGEANT BRUCE ALDEN BROWN AF13872031

FOR

YOUR HISTORIC ENDEAVOR TO PRESERVE FREEDOM OF THE ROK AND THE FREE WORLD FOR 30 OF MORE CONSECUTIVE DAYS OF SERVICE IN KOREA FROM 1968 THRU 1969

NATIONAL DEFENSE AUTHORIZATION ACT (PL 107-314) SIGNED BY PRESIDENT GEORGE W. BUSH ON DECEMBER 2, 2002

Capt Laurence B. McMurtry
Executive Officer

Donald Rumsfeld
Secretary of Defense

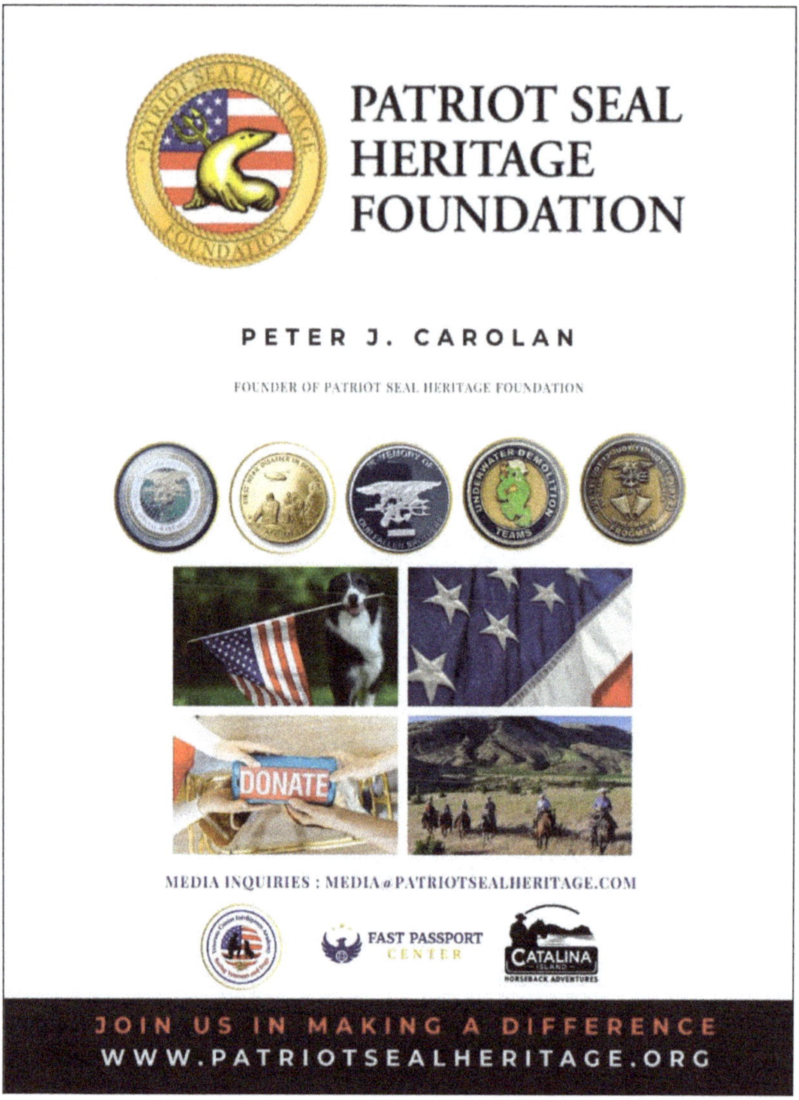
PATRIOT SEAL HERITAGE FOUNDATION
PETER J. CAROLAN
FOUNDER OF PATRIOT SEAL HERITAGE FOUNDATION
DONATE
MEDIA INQUIRIES : MEDIA@PATRIOTSEALHERITAGE.COM
FAST PASSPORT CENTER
CATALINA
JOIN US IN MAKING A DIFFERENCE
WWW.PATRIOTSEALHERITAGE.ORG

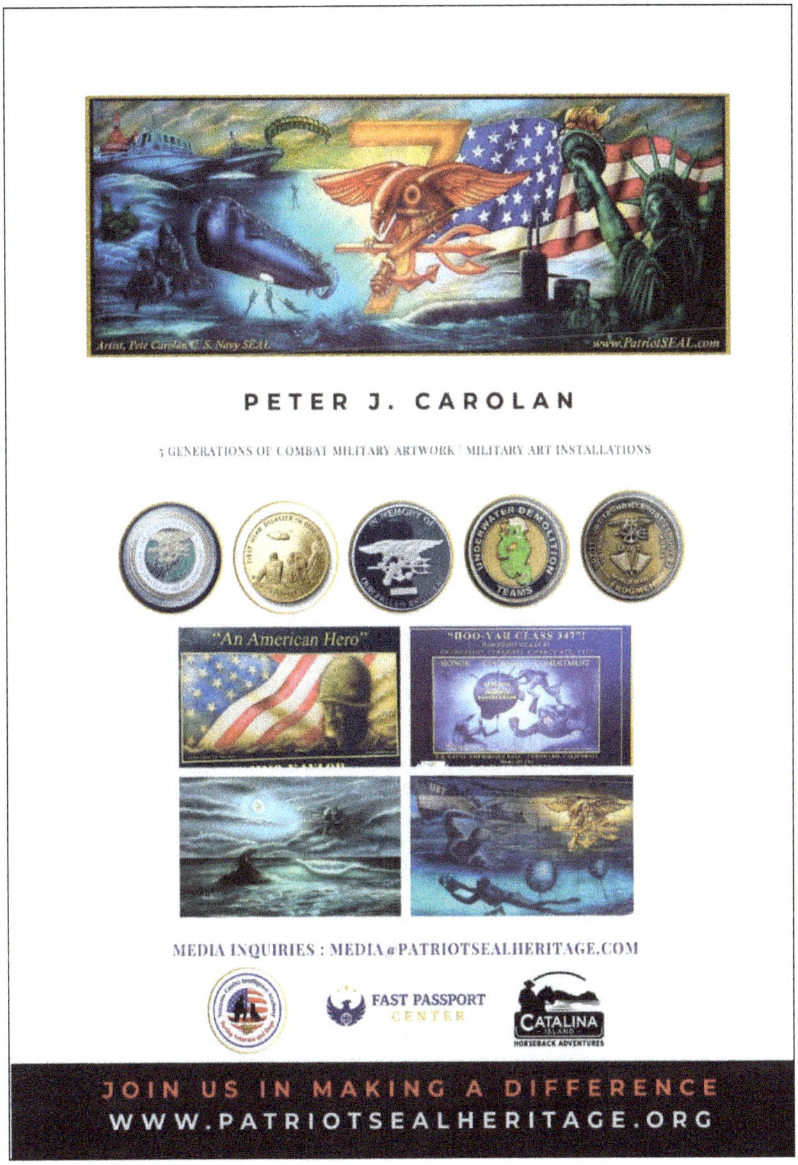

Artist, Pete Carolan U.S. Navy SEAL
www.PatriotSEAL.com
PETER J. CAROLAN
5 GENERATIONS OF COMBAT MILITARY ARTWORK | MILITARY ART INSTALLATIONS
UNDERWATER DEMOLITION TEAMS
FROGMEN
"An American Hero"
"HOO-YAH CLASS 347"
MEDIA INQUIRIES : MEDIA@PATRIOTSEALHERITAGE.COM
FAST PASSPORT CENTER
CATALINA ISLAND HORSEBACK ADVENTURES
JOIN US IN MAKING A DIFFERENCE
WWW.PATRIOTSEALHERITAGE.ORG

ABOUT THE AUTHOR

Dr. (h.c.) Lisa Ray is a literary agent, ghostwriter, and publisher with more than two decades of experience specializing in legacy storytelling, military history preservation, and mission-driven publishing. She is the Founder of Golden Ray Publishing, a publishing and media house dedicated to transforming lived experience into enduring narrative and cultural legacy under the guiding philosophy: *"Everyone has a story, what is yours?"*

Her work is distinguished by a deeply immersive, interview-based method, where she conducts extensive firsthand history interviews with veterans, leaders, and contributors to capture authentic voice, memory, and perspective. These interviews form the foundation of her storytelling approach, elevating each project beyond traditional biography into a curated historical archive.

In addition to publishing, Lisa curates visual and narrative installations inspired by her work, developing book-based art and exhibition pieces designed for museums, libraries, and traveling art shows. These curated works extend the life of each story beyond the page, preserving them as cultural artifacts that honor service, sacrifice, and legacy.

Her notable publishing work includes *MOY BELL: The Only Easy Day Was Yesterday,* the official memoir of Terry "Mother" Moy of SEAL Team One. Following the publication of the book, a replica of *MOY BELL,* donated by the Moy family, was placed in the San Diego Navy SEAL Museum in recognition of the work's contribution to preserving Naval Special Warfare history.

Lisa's current project, *The Painted Frog,* continues this immersive storytelling model through extensive interviews and archival collaboration, weaving together voices from the UDT and Navy SEAL community. The project also serves a larger mission: to support and fund initiatives through the Patriot SEAL Heritage Foundation (PSHF), advancing veteran wellness programs, equine and K9 therapy, and trauma recovery services for those who have served.

Lisa now serves as the Media Director for Patriot SEAL Heritage Foundation. Her work telling the ongoing military legacy story through the curated art of Peter J. Carolan. His works can be found in National Museums and will tour National libraries, art festivals, and exhibitions.

Dr. (h.c.) Lisa Ray

Through Golden Ray Publishing, Ms. Ray continues to build a platform for all where storytelling, art, and mission intersect—creating works that do not simply document history, but preserve it, show it, and actively use it to serve the communities that shaped it.

For more information or to connect please visit:
www.goldenraybooks.com

To learn more about the mission of the Patriot SEAL Heritage Foundation, including volunteer opportunities, ways to support its programs, and additional information, simply scan the QR code.

The QR code also provides access to Pete "The Pirate" Carolan's legacy collection, where you can purchase his artwork, books, commemorative merchandise, and official Patriot SEAL Heritage Foundation apparel. Every purchase helps preserve Pete's remarkable legacy while supporting the Foundation's ongoing mission to serve veterans and their families.

www.ingramcontent.com/pod-product-compliance
Lightning Source LLC
LaVergne TN
LVHW051921051126
851857LV00021B/2093

* 9 7 9 8 9 9 7 3 7 3 6 0 3 *